Minimum Economic Recovery Standards

Third Edition

Minimum Economic Recovery Standards

Third Edition

The SEEP Network

Practical Action Publishing Ltd
25 Albert Street, Rugby,
Warwickshire, CV21 2SD, UK
www.practicalactionpublishing.com

© The SEEP Network, 2017

Second edition published by Practical Action Publishing 2013
Third edition published by Practical Action Publishing 2017

ISBN 978-1-85339-957-2 Paperback
ISBN 978-1-78044-670-7 Library Ebook
ISBN 978-1-78044-957-9 Ebook

All rights reserved. No part of this publication may be reprinted or reproduced or utilized in any form or by any digital, electronic, mechanical, or other means, now known or hereafter invented, including photocopying and recording, or in any information storage or retrieval system, without the written permission of the publishers.

A catalogue record for this book is available from the British Library.

Citation: SEEP (2017) Minimum Economic Recovery Standards, Third Edition, Washington D.C., the SEEP Network and Rugby, UK, Practical Action Publishing <http://dx.doi.org/10.3362/9781780446707> Sections of this publication may be copied or adapted to meet local needs without permission from the SEEP Network, provided that the parts copied are distributed for free or at cost – not for profit. Please credit Minimum Economic Recovery Standards and the SEEP Network for those sections excerpted.

To access this publication online, visit www.mershandbook.org

This study is made possible by the generous support of the American people through the United States Agency for International Development (USAID). The contents are the respon sibility of the SEEP Network and do not necessarily reflect the views of USAID or the United States Government. This initiative is carried out as part of the AED FIELD-Support mechanism. For more informa tion, please visit www.microlinks.org/field.

Since 1974, Practical Action Publishing has published and disseminated books and information in support of international development work throughout the world. All print editions are produced and distributed via ethical and sustainable print on demand global facilities.

Practical Action Publishing is a trading name of Practical Action Publishing Ltd (Company Reg. No. 01159018 | VAT 880 9924 76). All profits are covenanted back to its parent group, Practical Action (Charity Reg. No. 247257).

The manufacturer's authorised representative in the EU for product safety is Lightning Source France, 1 Av. Johannes Gutenberg, 78310 Maurepas, France. compliance@lightningsource.fr

Contents

Using the Standards	vi
A Quick Look Inside the Standards	1
1 Core Standards	8
2 Assessment and Analysis Standards	38
3 Enterprise and Market Systems Development Standards	64
4 Asset Distribution Standards	92
5 Financial Services Standards	116
6 Employment Standards	140
Annex: Market-linked Tools and Frameworks for Assessments	156
Glossary	162
Standards Development Task Force	178

Using the Standards

This resource is not a 'how to' for implementing economic programs in humanitarian contexts. Its intent is to provide the reader with guidance on what good programming looks like and what to consider when you are planning your activities.

You can read each section separately or in sequence. Each section contains cross-references to other chapters or sections that may also be relevant, because many of the Standards or actions are linked. Use the figure on page 1 to guide you.

This book will be most useful to field practitioners and humanitarians implementing programs immediately after a crisis. Donors, governments, private-sector actors, proposal writers, and operational staff may also find it a helpful reference point when designing or reviewing project activities.

There will always be a tension between universal standards and the ability to apply them in the moment. Each context is different, and local conditions may make it impossible to meet these standards. This book provides the reader with an understanding of the final results that implementers should be working towards.

A Quick Look Inside the Standards

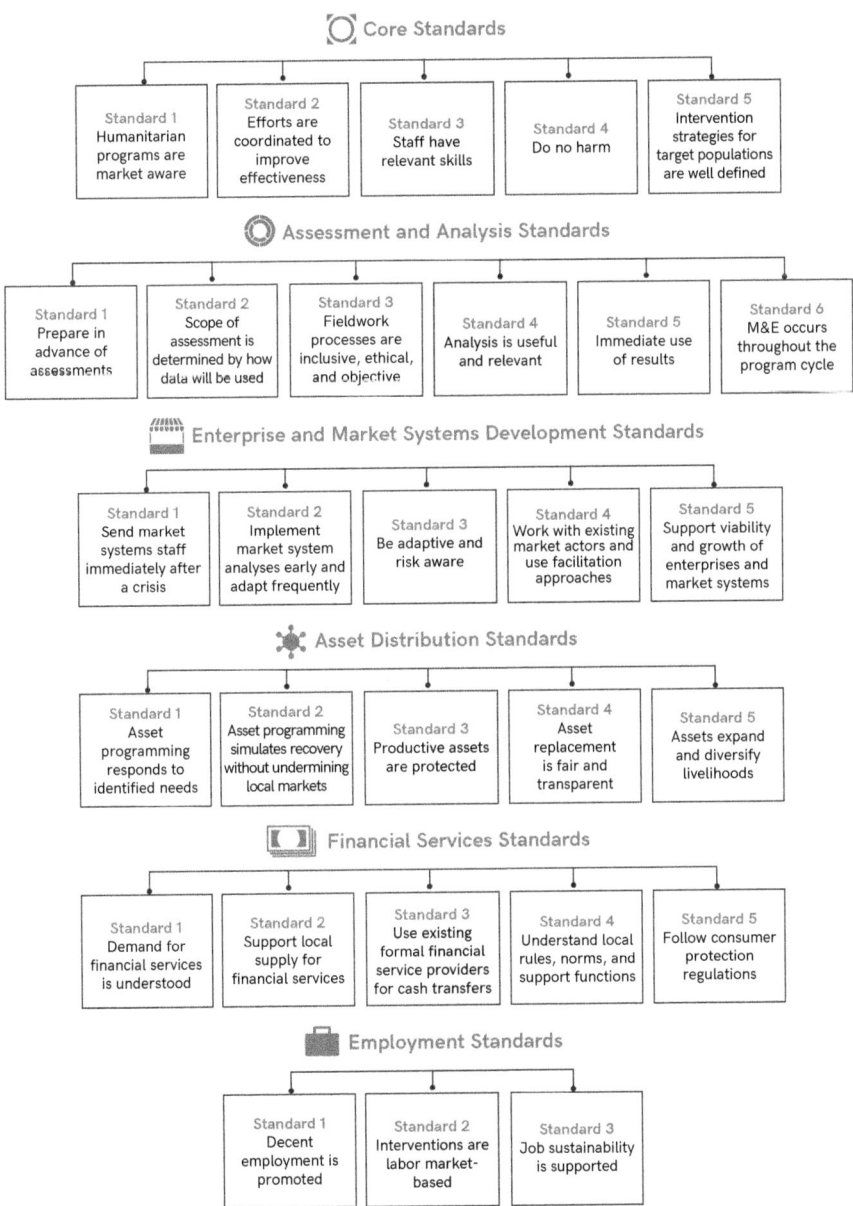

Who should use the MERS?

The Minimum Economic Recovery Standards *(MERS)* should be used by anyone planning or implementing economic or livelihood programs in a humanitarian context. It will also be useful for operational staff procuring and/or providing large volumes of goods for a specific area (such as non-food items distributions) to understand how to avoid a negative impact on the market. Donors, governments, private-sector actors, proposal writers, and evaluation staff will find it a useful resource for designing or reviewing project activities.

When should these Standards be applied?

As often as possible.

The Standards are designed to be used pre-crisis, in the earliest days of response, through recovery, to the beginning of longer term development. They are helpful any time you are interacting with a market – whether the response is intended to be market-neutral, market-aware, or market-integrated. They can be used for any market and for programs where economic or livelihood outcomes are not the primary focus of the activities.

Cross-cutting issues and specific target groups

In revising the *MERS*, care has been taken to address issues that are relevant across the Standards. These cross-cutting issues are: 1) gender, 2) disability, 3) preparedness, 4) resilience, 5) protection, and 6) the environment. They have been incorporated into the relevant sections of each chapter, rather than being dealt with in parallel. This book cannot address all cross-cutting issues comprehensively, but it recognizes their importance and makes links to other partner standards and resources where more information is available.

One point to note for those less familiar with economic programming, is that it is important to consider potential beneficiaries who would not normally be considered 'vulnerable' when targeting. Because they are not vulnerable, they are often the sole providers for their families or are able to hire others who are vulnerable. They can be part of the solution, reaching those most in need using existing community structures.

How to read the MERS: the difference between Standards, Indicators, Key Actions, and Guidance Notes

Each chapter presents a set of *Standards*, with *Key actions*, *Key indicators*, and *Guidance notes* for each standard.

The *Standards* are qualitative in nature: they are meant to be universal and applicable in any environment. They are the benchmark by which the quality of a set of activities can be judged. *Key actions* are the tasks that may be done by practitioners in order to meet the minimum standards. Note, however, that simply because a key action is taken does not mean the standard is automatically met. *Key indicators* are 'signals' that show whether a minimum standard has been met. They provide a way of measuring and communicating processes and results of key actions, and can be quantitative or qualitative. *Guidance notes* provide specific points to consider when applying the minimum standards, key actions, and key indicators in different situations. They provide direction on how to tackle practical difficulties or advice on priority issues.

A short history of the MERS, Sphere and the Humanitarian Standards Partnership

What is Sphere? The Sphere Project and its Handbook are well known for encouraging quality and accountability in humanitarian response. Initiated in 1997 by a group of humanitarian non-governmental organizations (NGOs) and the International Red Cross and Red Crescent Movement, the aim was to improve the quality of their actions during disaster response and to be held accountable for them. Sphere's philosophy is based on two core beliefs: first, that those affected by disaster or conflict have a right to life with dignity and, therefore, a right to assistance; and second, that all possible steps should be taken to alleviate human suffering arising out of disaster or conflict. Striving to support these two core beliefs, the Sphere Project framed a Humanitarian Charter and identified a set of minimum standards in key life-saving sectors which are now reflected in the Handbook.

In 2007 a group of practitioners, members of The SEEP Network, recognized the need to extend Sphere's guidance to economic programs taking place in humanitarian contexts. Noticing that often opportunities were missed or programs were poorly implemented, this group sought to explore emerging best practice and outline a vision for consistent and technically sound interventions. With the support of USAID's Office of U.S. Foreign Disaster Assistance, the *Minimum Economic Recovery Standards (MERS)* were created out of a collaborative space where practitioners built a shared vision for improved programming. The *MERS* have undergone two major collaborative revisions. This is the third edition, representing the work of hundreds of practitioners and thought-leaders over the last 10 years.

Sphere has now recognized four Partner Standards in addition to the *MERS: INEE Minimum Standards for Education: Preparedness, Response, and Recovery; Livestock Emergency Guidelines and Standards (LEGS); Minimum Standards for Child Protection in Humanitarian Action (CPMS)*; and the *Minimum Standard for Market Analysis (MiSMA)*.

The Humanitarian Standards Partnership (HSP), which began in 2015, grew out of the Sphere Companionship model, and promotes complementarity and coherence among technical standards. The HSP draws together the why, how, and what of humanitarian work and encompasses: *The Humanitarian Charter*, providing the ethical and legal backdrop to humanitarian response; *Protection Principles*, which set out how to protect people from violence, avoid causing harm, ensure access to impartial assistance, and assist with recovery from abuse; *The Core Humanitarian Standard*, which describes the essential elements of accountable, effective and high-quality humanitarian action; and the *Minimum Standards*, which provide universal benchmarks for assistance in shelter and settlement; water, sanitation and hygiene promotion; food security and nutrition; health; education; child protection; livestock; and economic recovery and market analysis.

This is how the Partner Standards complement the *MERS*:

- *INEE Minimum Standards* underline the importance of making sure that education related to livelihoods and employment – small business development, financial literacy, technical and vocational education and training – is provided to young men and women, particularly those from vulnerable groups who do not complete formal schooling. They encourage analysis of labor markets and collaboration with the economic and early recovery sectors to ensure

that the business skills learned are useful and programs are relevant for future employment.
- *LEGS* deepens the context of the *MERS* by providing benchmarks related to a critical productive asset – livestock – which so many communities rely heavily upon for their social and economic well-being. With climate change bringing more frequent and diverse types of disaster, *LEGS* also provides guidance for working with vulnerable livestock-dependent communities in fragile arid and semi-arid environments.
- *CPMS* provides a complementary set of agreed norms relating to child-protection work in humanitarian settings, including guidance on issues related to child labor, how the child protection and economic recovery sectors meet, and the release and reintegration of children from armed forces or groups.
- *MiSMA*, developed by the Cash Learning Partnership (CaLP), especially resonates with the *MERS* as both are built on the principle that market analysis should increase the quality of response and limit potential harm. The primary content of both sets of standards is aligned, with the main difference being *MiSMA* is intended to be used by humanitarian practitioners across sectors in an emergency, whereas the *MERS* goes into greater detail regarding the implementation of economic recovery activities and includes household economies and broader economic constraints in the market analysis. However, both can be used across all stages, from preparedness to early recovery.

The *MERS* third edition revision was a deeply collaborative year-long process involving over 90 organizations. Two write-shop events, and consultations in Geneva, Dakar, Panama City, New Delhi, Beirut, and London, involved more than 175 people in the drafting and reviewing process. A steering committee provided oversight and guidance, ensuring that the perspectives of multiple stakeholders were included and that the final document would be comprehensive, yet accessible.

To access the document online and for further resources and publications, visit www.mershandbook.org.

◯ Core Standards

- **Standard 1**
  Humanitarian programs are market aware

- **Standard 2**
  Efforts are coordinated to improve effectiveness

- **Standard 3**
  Staff have relevant skills

- **Standard 4**
  Do no harm

- **Standard 5**
  Intervention strategies for target populations are well defined

1 Core Standards

Markets play a critical role in how people survive, so understanding how they work during a crisis, and the best way for humanitarian responses to work with them, is essential. Markets are a physical and/or virtual space where people and businesses buy and sell goods and services; and response efforts take place within a country's economy as well as in its geographical area. Interventions need to be aware of market realities and how market systems link together the actors, governance and power dynamics, and formal and informal spaces where individuals, households, and businesses of all sizes come together.

Crises certainly affect markets, but not always in predictable ways. Crises can disrupt specific activities and relationships within a market or cause markets to fail completely. As a result, individuals, households, and businesses may be forced to take actions that undermine their current well-being and future viability. For households, this may include eating less, reducing spending on medical care and other essentials, removing children from school, and selling productive assets, such as livestock. Businesses may be forced to delay maintenance and investments, sell off equipment, and lay off workers. It is important to remember that some markets grow during and after a crisis, and some even thrive in this situation.

The *MERS* is a partner to the *Sphere Handbook* and as such shares the same foundations of quality and accountability for humanitarian standards. Specifically, the *MERS* Core Standards are harmonized with and emulate the Core Humanitarian Standards (CHS), adapted specifically for those working in economic recovery. The aim of linking the *MERS* and CHS is to establish stronger accountability to affected populations.

The *MERS* Core Standards help to ensure that programs meet the most basic responsibilities of economic recovery activities, and that interventions support opportunities for people to earn an income – via wage employment or self-employment – and to rebuild their lives, on their own terms, with dignity.

Core Standard 1
Humanitarian programs are market aware

Program design and implementation decisions consider context, market system dynamics, and communities. Market systems programming begins with the needs of the targeted groups.

Key Actions

- Learn which markets the program may interact with by speaking to procurement, logistics, and other operational teams in addition to the program staff.
- Determine if market assessment or analysis has taken place. If not, conduct one to develop interventions based on the needs of the poulation and sustainability of activities. Include a determination of economic viability, through a cost-benefit analysis and/or feasibility study (see also *Assessment and Analysis Standards*).
- Use the analyses to decide on your program activities. Consider the right level of market intervention and the views of a variety of stakeholders, including the community. Utilize existing market mechanisms, unless there is a compelling reason not to.
- Consult with relevant public and private-sector stakeholders to work in partnership.
- Establish monitoring systems to collect and analyze information on the market and the impact of the program. Feed this information (including social impacts) into project learning for ongoing improvements.
- Communicate the intent of economic activities with staff, partners, and community members, and be clear about who they can serve.
- Engage the targeted population throughout the program cycle.

Key Indicators

- Interventions invest in economically viable activities that target stable or growing markets.
- Interventions do not negatively distort markets.
- Interventions should be appropriate and relevant, with a clear logic for how they will meet people's needs and support capacities.
- Interventions have a monitoring system in place to enable regular and ongoing monitoring of changes (e.g. market prices, risks to market actors). Programs are regularly adjusted based on changing market and social conditions.

Guidance Notes

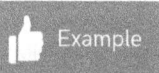 **Market systems**

Economic recovery should consider interventions that work at multiple points across a market, from input suppliers to producers, end markets, and policymakers. Programs that work at only one level and do not recognize these interconnections risk missing opportunities and creating market distortions with social consequences. Interventions may require a wide range of activities to have the greatest impact.

> **Example**
>
> Residents of camps for internally displaced persons (IDPs) in northern Uganda have prior experience in cotton production, but no longer have the resources to begin to grow cotton again. Dunavant, the regional cotton processor, has a supply gap of raw cotton and seeks suppliers. Rather than supporting these displaced people directly by distributing cotton seeds, the recovery organizations work with Dunavant to resolve significant barriers facing cotton production in the camps, such as collection of cotton within the safety of the camp and access to equipment to clear and plow land. By working with both the beneficiaries and the private firm that will buy their produce, the project is able to create long-term livelihood opportunities for many of these displaced people, which seed distribution alone would not achieve.

> **! Cautionary tale**
>
> A relief organization had been providing water trucking regularly for years in a drought-prone area, assuming that access to water was a supply issue for poor households. A market assessment was done and the team learned that the water market could meet community needs; the relief organization had been running a parallel system. The actual problem had been purchasing power, as the cost of water increased during droughts. The response shifted to supporting a community-level trading entity that would procure from local water truckers and manage distribution through a voucher system to ensure targeted distribution to vulnerable groups. Coverage of targeted groups increased significantly and the approach was adopted as the local government's preferred modality.

❷ Viability

Choosing the right program activities depends on understanding the markets in which enterprises and households operate. Economic recovery programs should orient enterprises and households to markets that are growing, stable, or have unmet demand in order to provide opportunities for employment and/or increased income to sustain livelihoods. Shrinking or non-competitive markets are ultimately not sustainable. Assistance that pushes people to stay in these markets will undermine their livelihoods in the long run. Additionally, non-competitive markets have fewer incentives that encourage entrepreneurs to invest, adopt new technologies, or benefit from program activities. This limits program effectiveness and reduces the ultimate goal of providing viable livelihood opportunities in communities affected by crisis. Understanding the affected market systems and marketplaces is a critical step and is covered in the *Assessment and Analysis Standards*, as well as the *MiSMA*.

On the other hand, economic recovery activities should target groups or individuals that are capable of sustaining and expanding their economic activity into the future. Should vulnerable groups lack the ability to sustain an activity, they are best served if, in addition to livelihood support, they receive cash transfers and other social safety-net interventions as well as capacity-building to build skills for the future. Such supplemental interventions should take into account the specific needs, capacities, and risks of sub-groups (e.g. women, people with disabilities, non-conforming gender identities and sexual orientations) to ensure inclusion, protection, and efficacy.

> **⚠ Cautionary tale**
>
> An agency conducts a livelihoods assessment and finds that many people are interested in raising livestock as an economic activity. The program provides households with cattle, sheep, and goats with the plan that they will sell the animals' offspring as a way to earn income. However, the agency did not consult the community deeply enough to understand that most households did not have the resources to care for the animals, lacking access to reliable feed sources and affordable veterinary services. As a result, many households found it easier to sell the animals immediately, without gaining any value from them by fattening them or collecting milk. Other households lost their animals to disease.

❸ Market distortion

Despite the best intentions, many emergency or development interventions can create market distortions. Market distortions include any unintended results that negatively affect a market system, ranging from extreme price fluctuations to the physical destruction of a market. It is the responsibility of those intervening in crisis situations to ensure that their interventions do not replace local products and actors, or otherwise create harmful distortions. Interventions should create positive effects beyond their intended economic targets, for example outcomes that empower women.

The primary objective of activities is to help local markets recover and support them to serve affected communities; however, caution should be taken with any activities that have the potential to distort markets or adversely affect communities, whether in the immediate or long term. Interventions that include local procurement and support local businesses can be designed to mitigate the risk of market distortion as well as help ensure the goods are culturally appropriate and meet local tastes (see also *Asset Distribution Standards*). Collecting and analyzing market data in a timely manner during the interventions can ensure local procurement and other activities do not create market distortions. One market distortion warranting special attention is corruption. While corruption exists in many markets, not just post-crisis markets, it is important that programs are aware of it and take proactive measures against it, rather than reinforce or allow it to flourish. This can also help reduce potential conflict.

> **Cautionary tale**
>
> After the Indian Ocean tsunami, many organizations promoted cash-for-work programs in an effort to inject cash into the economies and help households meet basic livelihood needs. However, many of these programs provided daily wage rates that were much higher than those earned by local farmers in their normal farm activities. Many households stopped farming or other small-scale production in order to benefit from short-term cash-for-work programs. This created a negative impact on the availability of local food and other locally produced items, and potentially had a longer range harmful impact on agricultural production. A deeper market and social analysis could have provided better targeting of potential workers, by recommending daily labor rates that did not create disincentives to continue local production.

④ Responding to changing market conditions

Markets are dynamic, particularly in crisis environments. Ongoing monitoring of the market system and the targeted enterprises or households will help identify emerging opportunities or constraints. Regular monitoring will also help interventions determine how best to adjust project investments (such as time and funding) for the greatest impact. Effective strategies can range from tracking changes in the availability of services and inputs that are critical to small farmers, to local price monitoring and meetings with regional wholesalers, to more complicated tracking of regional and international commodity prices and trends (see also *Assessment and Analysis Standard 1*).

Core Standard 2
Efforts are coordinated to improve effectiveness

For maximum efficiency, coverage, and effectiveness, interventions are planned and implemented in coordination with the relevant authorities, humanitarian agencies, civil society organizations, and private-sector actors. Coordination is internal and external.

Key Actions

- Determine if an analysis of the stakeholder environment in the affected area has been done. The analysis may include transporters, government actors, producer cooperatives, labor unions, and warehouse collectives. If not, consider conducting one to better understand who should be included in coordination efforts, looking at their capacities, power dynamics, and which actors are marginalized or excluded.
- Review the market regulatory framework, if it exists.
- Be informed of the responsibilities, objectives, and coordination role of government authorities and other relevant coordination groups.
- From the outset, participate in existing coordination meetings with local, national, and international actors, and avoid creating new structures where possible. Use these groups to undertake joint assessments, disseminate findings and other relevant information, and/or formulate intervention strategies and programs.
- Provide information about the agency's mandate, objectives, and economic recovery programs to the relevant coordination bodies and local stakeholders.
- Look at the enabling environment, government policies, and program objectives to determine if an advocacy strategy is needed to achieve programming results. Collaborate with other implementing agencies to strengthen advocacy on critical issues.
- Clarify agency practice regarding coordination and partnerships with the private sector and other actors in the response.

Key Indicators

- There is no duplication of interventions and programs among agencies in the same geographical or sectoral areas.
- Programs regularly exchange assessment reports and information with donors, implementing agencies, government stakeholders, local leaders, other humanitarian actors, and the private sector.
- Commitments made at coordination meetings are acted on and reported back in a timely manner.
- Organizations, programs, and projects that either cannot address identified needs or are unable to meet the minimum standards make gaps known, so that others may assist.
- An organization's response strategy reflects the capacity and plans of other humanitarian agencies, civil society organizations, and relevant authorities.
- Asset transfers and distributions are coordinated, sequenced, and aligned with the local economy to avoid responses undermining each other.
- An information-sharing mechanism between stakeholders is in place.

Guidance Notes

1 Coordination mechanisms

Uncoordinated responses lead to duplication, inefficiency, and potential conflicts in project strategies and interventions. This is particularly critical in economic recovery programs where different organizations' programs may undermine each other if they do not coordinate. An example of this is where agencies in the same geographical area or location provide grants and loans to the same target group for the same purposes, but at different rates and with different conditions. Lack of coordination may also burden disaster-affected people if multiple teams demand the same information for market assessments, when this information could instead be shared across agencies. Collaboration optimizes resources: a coordinated effort by communities, host governments, donors, and humanitarian agencies with different mandates and expertise maximizes coverage and quality.

❷ Coordination roles

It is the primary role and responsibility of the government of the affected state to respond to and coordinate humanitarian responses from assisting organizations. Humanitarian agencies play an essential role by supporting governments and respecting their coordination function. In some contexts, however, government authorities (and some civil society groups) may themselves be responsible for abuse and violations, or their assistance may not be impartial. In this context, a response coordinated with parties involved in the conflict may be inappropriate. Where the state is willing, but lacks capacity, humanitarian agencies should assist the state in fulfilling their responsibilities. In these contexts, sharing information across all sectors, as rapidly as possible, will enable agencies to respond to the needs of the affected population more quickly and effectively. Common forums for international NGOs to share such information include UN-led groups, such as the Office for the Coordination of Humanitarian Affairs (OCHA) and the Humanitarian Information Center, and cluster meetings for UN-declared emergencies. Coordination mechanisms may be monthly or quarterly meetings, an email listserv, the 4Ws (who, what, when, where) or an NGO forum.

❸ Transparency on targeting

Tensions can be high in crisis and post-crisis situations. Efforts must be made to effectively and openly communicate with all stakeholders. This can be done through transparent mechanisms, such as community meetings or local committees. Information about programs, decisions, and participation opportunities and criteria should be shared with all those affected by the crisis. Sharing information helps reduce misunderstandings, particularly when the program provides resources to only some people or provides a service that is new to the community.

④ Transparency on approach

Organizations working with the private sector should articulate their policies clearly to communities and other stakeholders. The needs of those affected by crisis are significant, and stakeholders may not understand or perceive how partnerships with the private sector can address the livelihood needs of those most affected by the crisis. Transparency and disclosure of why organizations partner with the private sector and other local actors, including commitments and potential gains, will reduce misconceptions of different partners' roles in the program. Humanitarian organizations working with private-sector actors should build on the work of the task force on Protection from Sexual Exploitation and Abuse (PSEA) and enforce data protection that protects program participants. Protective measures should be formalized within standard operating procedures and contracts and regularly monitored.

> In one crisis-prone area, the local government was very resistant to the market systems facilitation approach being implemented by a humanitarian agency. They wanted direct aid handouts to households because this was familiar to them (and easier for them to control), despite the fact it was causing food insecurity in the area, as very few businesses could sustain the sales of bulk food products. Through intensive advocacy to government officials and training of junior government staff on the *MERS*, local authorities came to understand the wider potential benefits of the approach and dropped their objections.

⑤ Making gaps known

When gaps in programming are identified and shared with other responding agencies, agencies with the appropriate technical expertise and/or spare capacity can more easily fill the gap. Timely information sharing about project locations, involvement of local partners, and emerging needs should be promptly communicated to the appropriate coordination bodies. The response coordination could be done through existing mechanisms (such as cluster or interagency) and aimed at identifying gaps, resources, and new partnerships.

 Example

A humanitarian relief agency notices a high demand for credit and savings among its target population, but the agency does not have the technical capacity to provide these services. Instead, it helps introduce staff from a local microfinance institution to community leaders in the target areas, and lets other agencies know of the financial service demand so they can begin exploring the best way to meet these needs.

6 Pricing, wage setting, and valuation of transfers

For interventions where assets are given to individuals and targeted groups (such as cash transfer programs, cash for work, distribution of equipment, and vouchers), the value of the distribution should be in line with Sphere standards and based on an analysis of current market prices and household needs. These values should be openly coordinated among donors and implementing agencies, and in line with government policy where required, so that there are no distortions in terms of prices or crowding out of existing private-sector providers. Equitable principles should also be followed in setting labor rates. Setting cash-for-work wages must take into consideration the local labor market in order to avoid increasing economic disparities and their social consequences, and to prevent 'poaching' or luring away workers from lower paying, but more sustainable, work. (See also *Asset Distribution Standards* and *Employment Standards*.)

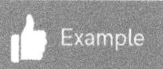 Example

In post-earthquake Haiti, many organizations designed cash-for-work programs. Rather than each establishing a labor rate independently, the agencies worked together and with the government to establish a local daily labor rate that would not entice workers away from their regular employment (salaried work or self-employment activity), while still meeting the basic livelihood needs of those most affected by the crisis and in immediate need of cash assistance. This was then adjusted as needed for geographic and sectoral areas with different prevailing labor rates.

Core Standard 3
Staff have relevant skills

Programs are staffed by individuals who understand economic recovery principles and/or have access to technical assistance. Programs include capacity-building components to improve the relevant economic skills of staff.

Key Actions

- Develop human resource (HR) systems that enable the organization to access competent talent with relevant experience for economic recovery responses. When appropriate staff cannot be identified, consider partnering with an agency that has the required skill, rather than implementing directly.
- Ensure mechanisms exist for networking and knowledge-sharing so lessons from each crisis environment can be learned by others in the organization.
- Budget time and resources for staff training and professional development. Ensure that staff get training on protection and child protection, gender, inclusion, appropriate codes of conduct, and other relevant topics on an ongoing basis.
- Develop staff assessment and management systems that promote accountability for results among staff.
- Encourage managers to lead their teams in ways that are adaptive, and use creativity to respond to the changing humanitarian context.
- Advocate to promote the *MERS* within government and bilateral and multilateral funding agencies. Donors play a major role in funding crisis recovery programs; their sensitivity to the *MERS* is extremely important.

Key Indicators

- Staff working on economic recovery programs have the relevant technical qualifications; knowledge of local economic activities, cultures and customs, and conflict dynamics; and/or previous economic recovery experience. These skills should complement those of others on the team to add value.
- Technical and managerial staff are provided with the necessary training, resources, and logistical support to fulfil their responsibilities.
- Managers are accountable for achieving program objectives and adhering to the *Sphere Handbook* guidelines, relevant companion standards, the *MERS*, and their agency's economic recovery guidelines.
- A staff code of good behavior and practice (including child safeguarding and sexual exploitation and abuse policies) is well publicized and understood by staff, partners, and affected communities. Safe, transparent, and confidential reporting procedures are in place, and complaints are followed up and addressed in a timely manner.

Guidance Notes

 **Core Humanitarian Standard Commitment 8**

This states that 'communities and people affected by crisis receive the assistance they require from competent and well-managed staff and volunteer'; therefore, the commitment requires organizations to support staff to do their jobs effectively and to treat staff fairly and equitably.

❷ Relevant technical skills

Managers of economic recovery programs should have experience in designing and implementing market-driven economic recovery programs in rural or urban settings, and be knowledgeable about market dynamics, value chains, supply and demand, and so on. Technical expertise should be augmented with relevant short-term support when expertise cannot be found locally. In conflict settings, staff members should have experience analyzing and managing efforts that seek to mitigate and manage conflict. If this experience is not available among the long-term staff, a conflict specialist should be brought in at critical points in the program, especially during assessment and program design, and for periodic monitoring. National staff should be recruited wherever local capacities exist. It may not be possible for humanitarian organizations to cover this diverse skill set in-house, therefore strategic collaboration with knowledge organizations, short-term consulting opportunities, or even advisory groups, could be explored.

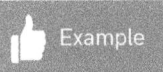

In Uganda's RAIN program, recruitment interviews included problem-solving scenarios or trips to the market to see how candidates analyzed context in real time. The contextual knowledge and analytical skills of these team members were invaluable in several program pilots.

❸ Staff training and capacity-building

Staff should receive basic training in the methods employed by the economic recovery program, as well as general introductory training in the targeted sectors. Opportunities for on-the-job training, e-courses, mentoring, and attendance in higher level economic development workshops should be made available to staff to build their program development and management skill sets. For staff that may be transitioning from distribution or relief projects, if their skills are relevant, specialized training should reinforce the importance of sustainability, the development of an appropriate exit strategy, and other good economic recovery practices.

Core Standard 4
Do no harm

The design, implementation, outputs, and environmental impacts of economic recovery interventions address or minimize potential harm, and do not exacerbate economic disparity, conflict, or protection risk, or undermine rights.

Key Actions

- Conduct a risk analysis to identify any ways that the intervention may increase the risk of exposing participants to increased danger or abuse, noting how these risks will be addressed and/or mitigated. Include environmental issues where possible and appropriate.
- Conduct ongoing monitoring of project operations, for example using a risk matrix, so that any unforeseen or new risks can be addressed quickly.
- As per CHS Commitment 5, 'communities and people affected by crisis have access to safe and responsive mechanisms to handle complaints', engage affected communities in the design, implementation, and monitoring of a complaints-handling process.
- Put in place systems, appropriate internal controls, and codes of conduct to reduce the potential for staff corruption, and transparent systems to reduce the potential for corruption with partners and government.
- Ensure humanitarian programming does not conflict with the local traditions or context, in particular when providing products or services to the population.
- Where possible, put in place mechanisms that promote peaceful coexistence between population groups in the implementation areas. Use dialogues related to economic recovery or resource use as a platform for deeper conversations.

Key Indicators

- The risk analysis is broken down by population group and considers specific vulnerabilities related to gender, age, disability, displacement, and social marginalization.
- Programs apply a 'do no harm lens' to selected market chains and enterprises to determine the wider social and environmental impacts of interventions and seek to proactively mitigate those identified.
- A risk matrix or contingency plan is used regularly to inform emergency programming.
- Interventions do not fuel divisions within the targeted communities but contribute to bringing people together and reducing tensions.
- Program participants are not subject to harassment, indecent working conditions, or increased safety and security risks as a result of their participation. Participants' safety and security should be enhanced as a result of their participation.
- Interventions are regularly monitored to ensure that no exploitative activities (such as corruption) are undertaken by staff, selected partners, and targeted enterprises; robust complaints mechanisms are designed, implemented, and monitored for all sub-populations in the consultation of affected communities and should cover programming, sexual exploitation and abuse, and other abuses of power.
- Programs seek to identify solutions to help people become less vulnerable to future crises.

Guidance Notes

 **Do no harm lens**

Economic recovery and market-based interventions can impact social power dynamics and potentially damage fragile relationships between groups, including men and women, displaced and host populations, or other self-identifying groups. Risk analyses and market assessments should specifically consider this dynamic to determine how program activities can reduce these risks. In mapping market relationships and power dynamics, all actors (input suppliers, producers, processors, traders, wholesalers, and retailers) should be included. Information should be gathered about the social networks in which they function and their traditional roles. Questions that can identify these ethnic/tribal, gender, or traditional roles should be included in questionnaires. As understanding grows, the power dynamics will be revealed, as well as which groups (men, women, children; ethnic, religious, caste, or tribal groups) traditionally or predominantly fill specific roles in the market system.

In some cases, a program that wants to assist a marginalized population can put market actors at risk if the social order is disrupted without the community's input. Tools such as network analysis can show which social networks exist, who is excluded or included, and why, so that interventions can be designed to enable more equitable and transparent relationships, and to leverage existing relationships and links. These dynamics will continue to evolve over the life of the intervention, therefore market assessment and gender and risk analysis should be done on a rolling basis and integrated into the program life cycle. A complaints and response mechanism provides an additional layer of feedback in which protection risks to the affected communities and individuals can be monitored and followed up outside the scope of the program cycle.

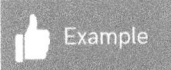 Example

As part of a country-wide initiative, the RAIN team conducted an internal gender analysis. They began tracking gender-disaggregated recruitment metrics against targets and proactively recruiting and promoting women. When a few of the newly recruited women on the team struggled to have their voices heard, the gender advisor, HR department, and the RAIN program director held a meeting to identify ways to resolve the problem. The program director focused on attracting women to the team and called on male gender champions within the team to support the women's integration. A year later, women made up 48 per cent of program staff, up from 22 per cent at the start of the program. Without deliberate efforts to support women's voices and retention, critical gender-focused shifts in program strategy would likely not have occurred. The project then extended this gender inclusion work to its private-sector partners by helping an important local wholesaler to increase its recruitment of female extension agents from none to 20 in two harvest seasons. These efforts helped the wholesaler increase the number of female contract farmers it registered and trained from 26 per cent to 48 per cent over the same period. This shift required dedicated resources: in addition to support from the country office's gender advisor, the team leveraged gender-focused grant funding in pushing forward the changes.

❷ Minimize negative impacts on the environment

Market development and post-crisis support of economic activity may put too many demands on the environment and potentially contribute to future food insecurity and other disaster risks. Traditional livelihood and income-generation activities often require natural resources as inputs: for example, water for farming, timber to make charcoal, or reeds for baskets. The steps in processing products need to be analyzed for potential negative impacts (for example, safe use and disposal of chemicals). In addition, decision-makers and program designers must carefully consider the role of natural-resource management in contributing to peace-building or, on the contrary, to fueling tensions and conflicts. Certain segments of the economy may be much more sensitive than others. The choice of interventions should be based on the results of analyses of potential negative environmental impacts, and the interventions should include methods for eliminating and/or minimizing negative impacts. Tools to assess these impacts include the Rapid Environment Impact Assessment (REA) as employed by CARE, UNHCR, and others.

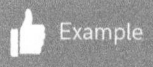
Example

Producing charcoal, which is an important fallback livelihood strategy in many places, leads to deforestation and pollution. Simply discouraging this activity on environmental grounds does not address the real economic need of families for the income it provides. Both the economic benefit and the environmental damage need to be weighed and balanced, with alternative solutions developed when possible.

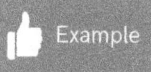
Example

A market program in Nepal encouraged the planting of fodder species on marginal land to mitigate soil erosion on slopes and improve dairy production in the area. The grasses retained soil and stabilized landslide-prone slopes – a clear disaster risk reduction activity – while increasing inputs to the growing dairy market.

❸ Conflict sensitivity

Decisions must be based on an accurate and up-to-date conflict analysis which comprehensively considers the root causes of the conflict, the profiles of conflict-affected groups, and the dynamics among stakeholders directly or indirectly involved in the conflict. Given that conflict settings are volatile, the situation and its trends must be regularly monitored. Consequent actions must not fuel tensions and divisions; instead, they must develop and protect the society 'connectors' from being undermined. Conflict sensitivity entails being aware of the impact of activities and implementation processes on increasing or reducing tensions among groups. This involves listening to the voices of the most vulnerable and marginalized; making decisions and designing programs to be as participatory, inclusive, and transparent as possible; avoiding an over-emphasis on certain groups, while progressively extending assistance to the community as a whole; addressing psychological trauma to facilitate reconciliation and remove resentment; establishing institutional mechanisms for timely and fair access to land and other assets as early as possible; and considering the impact of interventions at the regional and sub-regional levels.

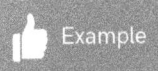 Example

In Myanmar, to achieve project goals, stakeholders from different sides of the conflict needed to collaborate. To increase collaboration, the project team focused the various parties on the overarching and shared ambition of meeting basic health needs, while managing the concerns and motivations of different actors and facilitating productive engagements. Activities such as joint training helped to build trust and personal connections.

4 Assess the exploitative potential of individuals

After a crisis, whether natural or human-made, there is often a gap in the governance that provides order in a society. Curbing illicit activities requires engagement with local communities and partnerships with non-economic programs, local government agencies, and multilateral donor agencies (see also *Employment Standards*).

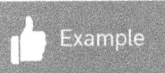 Example

An agency identifies a market opportunity for hand-knotted rugs in a post-crisis region. The project requires sources of natural dye, thread, weaving, trading, and exporting. Rug-producing households see an earning potential and remove their children from school to produce the rugs. The agency recognizes the problem at the outset and partners with local government, local schools, and some international donors, who vouchers reduce their household ex provide food vouchers for children who attend school. This gives the families an incentive to keep their children in school. The food vouchers reduce their household expenditure, which then enables families to hire outside laborers to assist in the rug production.

⑤ Political economy

In any market, there are entrenched interests, incentives, relationships, and dynamics. Any intervention that does not consider the existing incentives that drive the behaviors of market actors runs the risk of being ignored at best, and actively sabotaged at worst. It is therefore essential that the interests, incentives, relationships, and market dynamics are well understood and taken into account in the design phase of any intervention. For example, the top-down introduction of mobile money in an environment where the government is not fully supportive, or where it is not a development priority, will face serious challenges. While incentives can be altered over time through careful and intentional market facilitation, implementers must be aware of the starting point and have a strategy to address how to shift pre-existing incentives and dynamics.

⑥ Disaster risk reduction

In areas that experience frequent or recurrent crises, the development of resilient livelihood strategies is important to reduce the damage suffered from future shocks. While approaches can vary significantly by context, they may include:

- conducting a market assessment of key goods, natural capital, and services in a crisis-prone area, using tools such as the Pre-Crisis Market Analysis (PCMA) (see also Annex and *Assessment and Analysis Standards*)
- establishing and/or reinforcing formal and informal safety nets
- implementing Sensitive Ecosystems Inventories (SEI) initiatives that protect ecosystems
- carrying out projects, such as cash for work, that rebuild or protect community assets
- ensuring that assets and interventions provided can withstand future shocks (developing resilient market systems)
- providing training on income diversification and alternate skill development so households can rely on other sources of income if a future crisis harms one source (to increase their coping capacity).

In many cases, a root cause of disaster is a degraded environment. Policies and programs that promote climate- and environment-smart practices are often less costly, more effective, and more socially sustainable than infrastructure. When infrastructure-based Disaster Risk Recovery (DDR) activities are used, however, it is critical that they address environmental sustainability so future risks are not increased and neighboring communities are not adversely affected. It is critical that economic recovery interventions consider people's most likely methods of coping with a future shock, but these coping mechanisms should not be seen as more important than economic viability when designing programs.

After the 2009 earthquake in northwestern Pakistan, a system was set up to provide cash grants safely and securely via debit cards. Following the 2010 flooding in Pakistan, agencies could make use of and expand that same system, saving time and effort.

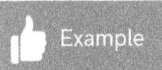

In Timor-Leste, one project used an integrated DRR and economic development model to demonstrate a more sustainable impact on risk reduction. In vulnerable communities, legumes or bananas were planted in fallow or marginal lands to improve soil fertility and as ground cover to protect from erosion, while also improving the food security of households and earned income through sales. Farmer groups that produced moderate amounts of legumes and bananas preferred to use them for household consumption and sell the excess to their neighbors to meet small expenditures. With food security a consistent problem for the village, the target farmers prioritized selling produce to their neighbors and the feeder program at the local school, rather than making more money by selling in the larger markets.

Core Standard 5
Intervention strategies for target populations are well defined

The intervention strategy is based on solid household and market analysis, and promotes the use of local resources and structures whenever possible, to help targeted households or enterprises reach the desired economic outcomes.

Outcomes for target groups may be achieved through a variety of strategies and partnerships, involving direct and indirect interventions. Indirect interventions (working through existing private-sector actors, for example) are likely to reach more households and be more sustainable because they use structures and networks that are already in place. There will be times when direct intervention is preferable for reasons related to capacity, geography, local politics, and so on. Nevertheless, it is preferable to use local structures and implement through facilitation approaches when possible, or at least to build capacity for local actors to implement the activity in the future.

Key Actions

- Determine if an assessment of the socioeconomic situation has been done. If not, implement one to better understand the vulnerabilities and capacities of the target population and its various sub-groups (such as women, girls, boys, men, people with disabilities, people of non-conforming sexual orientation and gender identity, and ethnic and religious minorities). The package of support offered should be informed by information from households and the mitigation strategies currently employed by communities that can be strengthened (see also *Assessment and Analysis Standards*).
- Conduct a market assessment to learn which points in the market system may be most in need of support, what networks and resources already exist, and what is likely to have the greatest impact on the target population and the intended outcomes (see also *Assessment and Analysis Standards*).

- Consider both direct and indirect intervention strategies when determining the most effective approach, and evaluate the risks of different intervention strategies. Identify ways to use local resources and structures, whether implementing directly or indirectly. Where indirect (facilitation) approaches can be used, the sustainability of the activity is likely to increase.
- If considering indirect approaches, work backwards from the targeted population to identify who is already providing goods or services (of any kind). Find out if those entities can adapt and become a partner in the response delivery.
- Use a logical framework, results chain, or other tool to describe the approach clearly. The approach is likely to change somewhat during implementation as the activities evolve, and the description of the approach should be updated.
- Ensure that affected communities can make informed choices relating to humanitarian goods and services.

Key Indicators

- Interventions use socioeconomic assessments, vulnerability analyses of target populations and market information to design support packages for traditionally excluded groups.
- The intervention approaches used are appropriate for achieving the desired outcomes in the given context.
- Market assessment and analysis have identified possible impacts and risks associated with different interventions for specific target groups, and program design, implementation, and monitoring account for these risks and include mitigation mechanisms.
- Program targeting and selection strategies include adequate risk mitigation and careful consideration of sociocultural and micropolitical factors.
- All programs clearly demonstrate and justify the linkage between the proposed intervention and the desired benefits for the target populations.
- Monitoring tools are developed to track the impact on the selected population.

Guidance Notes

1 Protection of dignity

The impacts of a crisis will be experienced differently by different individuals within a household. A good assessment will set the context for identifying the most appropriate means to achieve the desired impact on the intended individuals. Within any target population there is likely to be a wide range of capacities and vulnerabilities. Intervention strategies must identify and disaggregate risks by age, gender, and disability, and seek to address them in order to support all aspects of the dignity of targeted individuals.

2 Understanding the full context

Knowing where to place an intervention must go beyond identifying the needs of the intended recipients. Assessment tools should look at households, the economic contexts in which they operate, and the various market dynamics that affect the value chains in which they work. Such assessments help identify the feasible kinds of impact from any given intervention, as well as helping to clarify the potential risks that can occur as a result of an intervention. Economic activity creates environmental consequences, which must be understood and mitigated when possible (see **also** *Assessment and Analysis Standards*).

Example

> To increase the resilience of refugee farmers to cope with and adapt to the effects of climate change, an agency planned to increase access to irrigation water in a semi-arid area. This would allow the production of high-value crops not otherwise available in the surrounding markets. The agency conducted a market analysis to assess the potential of introducing an efficient irrigation system that would be available at the local market level and would not require high maintenance costs or management expertise. It consulted with refugees and involved water users' associations to oversee the management. This has facilitated the introduction of new technology and presented a model for government and donor investment in the area. New interests in processing and packaging were also created along value chains that had not previously been identified.

❸ Direct and indirect strategies

In many cases, beneficiaries have more to gain through indirect interventions at the market systems level than from direct engagement with individual households. Organizations must analyze the context, review local capacities and networks, and judge each situation based on the expected outcomes. Decisions should be based on the expected impact on targeted groups or areas, and may consider the potential for scale and sustainability.

Direct assistance can be the best strategy if the constraints identified are clearly at the household level, rather than related to broader access to finance or market issues. Indirect interventions may be more effective if the support required is long term and cannot be addressed by a one-off activity. A combination of approaches may also be required. In many circumstances, supporting target groups or partnerships may be insufficient to achieve the desired impact. Thus, complementing direct assistance with supporting interventions may more likely result in the desired outcomes.

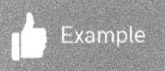

Women in a fishing industry in a flood-affected area were seeking to re-establish their business drying fish for local and regional consumption. Before the disaster, they dried fish on tarps on the ground, which resulted in contamination and a high moisture content, consequently limiting them to selling their dried fish in low-value markets. A local fish-processing company was interested in sourcing dried fish locally and had staff with experience in low-cost, effective techniques that could increase the value of the fish. However, the company did not see the women as valuable business partners. Direct assistance was provided to the women in business planning and group formation, which enabled them to link up with the company and access its technology. At the same time, assistance was provided to a local bank to develop loan products for small-scale fishing operations (men and women), which gave them access to working capital to scale up their businesses.

❹ Risk analysis considers stakeholders

Local context and culture play a significant role in defining how different people interact in the market, based on their gender, religion, ethnicity, age, social status, disability status, and a host of other defining characteristics. Power structures within families and societies create various opportunities and restrictions on an individual's access and opportunities within the market. The impact of these structures must be understood and the project must respond accordingly by taking assigned roles into account and promoting access and opportunities for all by seeking or reinforcing incremental change that ensures protection of all participants (see **also** *Core Standard 4*).

 Example

> In many countries, women represent the poorest segments of society. As such, many organizations have intervention strategies that reach women directly. Within some cultures, however, women's roles are clearly defined and changes to their roles are considered a threat to the society at large. Providing women access to loans to expand their businesses, for example, may be perceived as a threat by their husbands as it may shift the power dynamics in the household. In such circumstances, organizations may choose to work with village elders, husbands, or fathers – all those who are at risk of losing power – to obtain their buy-in and to demonstrate overall societal and household gains in order to secure acceptance. Deciding to serve both female and male clients may be another means of addressing this potential risk.

⑤ Demonstration of impact

All interventions, whether direct or indirect, should articulate how the target population will be affected and note the key assumptions. This may be particularly important with indirect intervention strategies to ensure that the assistance provides benefits to the intended groups and individuals (see also *Assessment and Analysis Standards*). Tools such as theories of change and results chains can help describe the assumptions and expected changes over the life of the project. If these project tools are visible to staff and updated regularly as assumptions change, the interventions are more likely to remain aligned with overall project goals.

Assessment and Analysis Standards

- **Standard 1**
  Prepare in advance of assessments

- **Standard 2**
  Scope of assessment is determined by how data will be used

- **Standard 3**
  Fieldwork processes are inclusive, ethical, and objective

- **Standard 4**
  Analysis is useful and relevant

- **Standard 5**
  Immediate use of results

- **Standard 6**
  M&E occurs throughout the program cycle

2 Assessment and Analysis Standards

Assessment and analysis are essential to good design and implementation of economic recovery programming. Analysis of market dynamics throughout the life of an intervention is necessary to ensure activities are having the intended outcome, and to identify opportunities for scaling as well as potential threats to sustainability. This chapter stresses the importance of ongoing program monitoring and evaluation (M&E), dissemination, use of results, and adaptation of interventions and strategies based on findings. Each set of standards in the *MERS* assumes use and comprehension of this chapter.

In this chapter, 'assessment' generally refers to investigation and study (using primary and secondary data) conducted before (and sometimes during) an economic recovery intervention. 'Evaluation' generally refers to reviews done at the end of the intervention that determine the program's performance and impact. Implementers should consider the purpose or goal of any assessment as well as the need to stage assessments during the project design and implementation processes. Data collection should be participatory, inclusive, cost-effective, and transparent. Information should be kept up to date and relevant, and data and reports should be made available in the public domain. Consider also the range of available (and evolving) tools for market assessment in the Annex that can help guide your assessment options, and note in particular the *MiSMA*, which is also a Sphere Companion Standard.

Assessment and Analysis Standard 1
Prepare in advance of assessments

Key elements are in place to conduct an assessment. When an emergency occurs, preparation for the assessment can build quickly on previous planning and existing resources.

Key Actions

- Identify high-risk countries or areas that are prone to crises. To be ready for humanitarian situations, be informed of existing assessments and use them in decision-making.
- Make a list of the key information that you need to know and the types of data needed.
- Identify existing information bases from the public sector or other reliable sources.
- If, after checking existing databases, you are unable to find relevant information, you may need to create a new database, ideally covering an extended period (e.g. time-series data).
- Determine your organizational readiness by identifying critical gaps in data, and designing mechanisms to compile and analyze any missing data on a regular basis.
- Compile up-to-date baseline information and generate mapping of relevant stakeholders, vulnerability, and potential shocks.
- Set up procedures, tools, and a roster of professionals to rapidly activate an assessment.
- In countries prone to crisis, strengthen the capacity of local assessment teams involving participation of multiple stakeholders.

Key Indicators

- Assessments are done prior to program interventions and regularly inform decision-making.
- Assessments build on existing data sources and are updated as needed. Assessment data should be disaggregated by age, gender, and disability and analyzed accordingly.
- Relevant baseline information is up to date and quickly accessible.
- Secondary data sources have been identified.
- Data collection tools are designed to gather necessary information and fill data gaps.
- A map of relevant stakeholders, vulnerability, and potential shocks is available.
- The necessary resources (capacity, financing, equipment) are ready to be mobilized.
- The capacity of assessment teams is sufficient to ensure a good-quality output.
- Information gathered during preparedness activities reduces the time spent preparing the assessment and conducting the analysis.

Guidance Notes

❶ Market-linked tools and frameworks for assessments

A wide variety of resources are available in the Annex. Teams should have tools ready which can be adapted for specific assessments and evaluations. Having a centralized place where these tools are kept, including examples of completed ones, which is accessible across the organization, can help field teams to select the right kind of tools and assessment staff to get comfortable with tools more quickly.

② Context analysis

Secondary data is information from existing reports and assessments (possibly done by other organizations). Primary data is information that will be gathered directly by the team. After reviewing information and context analyses that already exist, determine which primary data is likely to strengthen and update the analysis of the existing secondary data. Think about what is likely to change in the area of focus (for example, what type of crisis is most likely and what the risks are). In crisis contexts, it is usually more useful to focus on context data that is quickly changing, so being prepared for a crisis may involve gathering slower-changing data to use as a baseline when the crisis hits. Focus additional data gathering on information needed for decision-making.

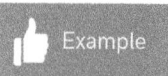
Example

> The South and Central Syria program's Humanitarian Access Team (HAT) is composed of five analysts who leverage their networks, including through the program's partners', to supply conflict analysis and forecasting across the response region. Early separation of the analytical team and the program team allowed space to incubate the analytical capacity, though this initially reduced the usefulness of the analysis. The two teams now collaborate more closely, and the HAT has begun expanding into political economy analysis and other areas.

③ Stakeholder analysis

Preparing a list of likely stakeholders can save time when a crisis hits. This list may need to be updated after the crisis, depending on what has changed. Remember that access to markets and livelihoods can vary according to age, gender, and disability so make sure all these groups are consulted (See also *Assessment and Analysis Standard 2*).

4 Assessment formats

These should be flexible and specific to both the hazard and the geographical context. The team should think about how much qualitative data is needed, (as compared to quantitative) and which assessment formats are likely to provide the right balance. Consider which technology (e.g. tablets, smart phones) is appropriate or helpful to data-gathering. Issues such as budgets and connectivity in survey areas can affect the final decision but technology can also help save time and avoid errors if assessment staff are trained properly. It is usually best to procure and get systems in place prior to an emergency. If staff are familiar with their use, non-paper systems can be very effective in a crisis context.

5 Staff training

Make adequate time for staff to be trained in the assessment formats, and test whether the tools used are gathering the desired information. Ensure that any enumerators understand the questions as well as any sensitivities that might exist around the questions. It is common for people to be reserved when discussing income or market success/failure. Groups that are disadvantaged in normal times (due to gender, age, and/or disability) may be even more reserved than others (see also *Assessment and Analysis Standard 3*).

> As part of an Emergency Market Mapping and Analysis (EMMA) training exercise in Vietnam, trainees surveyed the local market and built market maps for mosquito nets. An unexpected learning outcome from the training was that local companies could easily produce the quantities required for pre-positioning and emergency response. Several trainees made a commitment to immediately change their procurement practices; in the following season, two organizations procured their nets from local producers.

6. Logistics

Because market systems cover a broad geography that is sometimes beyond the immediate disaster zone, it is worth considering how the availability of transport will affect the scope of the assessment. Look at the make-up of teams and ensure women are involved appropriately, so that stakeholders who are not comfortable talking to men can also be included.

7. Funding for assessments and monitoring and evaluation (M&E)

The need for increased budgets for ongoing assessments is becoming better understood. Funding may be sourced from host-country governments, multilateral and bilateral donors, and some NGOs. Crisis contexts evolve quickly, so, wherever possible, write ongoing assessments into proposals to allow for more adaptive program response. Host-country governments and donors can also be good sources of information; many countries and local communities have early warning systems in place that can be important sources of information when preparing for assessments.

Assessment and Analysis Standard 2
Scope of assessment is determined by how data will be used

Decisions on how to use data are based on the specific situation and bring together critical information from key stakeholders.

Key Actions

- Define the overall scope and timeline of the assessment. Define the information requirements and set your key assessment questions based on the consideration that assessments are instruments for decision-making.
- Conduct stakeholder analysis early in the process to identify the appropriate stakeholders and their level of involvement in the assessment. Repeat and update as needed throughout the assessment process.
- Use evidence-based conceptual frameworks (e.g. sustainable livelihoods) to inform planning for assessments, analysis, and interpretation of data. Fit questions and methods to the plan for analysis.
- Make use of existing assessments and other secondary data sources before undertaking primary data collection. If this is not done, time and resources may be wasted. Review existing information, including pre-crisis data and desk studies, to avoid duplication, and determine what gaps in data exist or need to be updated through primary data collection.
- Before beginning an assessment, develop an assessment plan based on identified gaps and the knowledge necessary for effective decision-making and resource allocation. However, allow a good deal of flexibility once to use understanding of the context becomes clearer.
- Coordinate agencies and organizations in assessing specific sectors, technical areas, and communities to make the best use of resources and areas of expertise.

Key Indicators

- A clear assessment scope of work, informed by relevant factors, is developed first and an assessment plan is then produced.
- Critical information from key stakeholders is gathered and synthesized, including information about affected households' livelihoods, market systems, sociopolitical and conflict dynamics, and such considerations as gender, youth, and the environment.
- A stakeholder analysis identifies the key stakeholders and their level of involvement in the assessment.
- Relevant evidence-based conceptual framework(s) are used to guide assessment.
- Existing data sources are consulted to assess the availability of desired information.
- Assessments use a systemic approach that places economic recovery strategies within a wider context of market systems, economic trends, and political and socioeconomic institutions.
- Assessments are sensitive to ethnic, caste, gender, and wealth differences within populations. They measure the differences in economic opportunities between these groups and identify important existing or potential causes of conflict or marginalization.
- Assessments are participatory, conducted at appropriate times, incorporating seasonal calendars, security, market and labor trends, and other relevant social and economic factors, conditions, and trends.
- Assessments include a wide range of economic actors, including women, men, youth, people with disabilities, and elders; producers, traders, transporters, and consumers; local, regional, and national markets; and private and public market support functions.

Guidance Notes

❶ Assessment plan

The assessment plan should include: 1) assessment objectives, 2) planned use of data and findings, 3) conceptual frameworks to be used, 4) time and resources available, 5) what questions will be studied and what data or information will be collected, 6) how data will be collected (methods), and 7) who needs to be involved and where data will be collected.

❷ Stakeholder analysis

When conducting stakeholder analysis, make a list of potential stakeholders who have an interest or influence in the assessment. Ask: 1) what is their interest in the assessment, 2) what is their potential influence on the assessment, 3) what is their relationship with other stakeholders, and 4) what is their capacity or motivation for participating in activities that may be designed as a result of the assessment. Then ask: 1) who needs to be informed, 2) who needs to be consulted, and 3) who needs to be actively engaged at each step of the assessment process. The stakeholder analysis should consider involvement of affected communities, public and private institutions, UN agencies, and national and international NGOs.

❸ Conceptual frameworks

Once you have selected the relevant evidence-based conceptual framework, use it to develop assessment questions and ensure that you have gathered sufficient information to conduct an analysis. Conceptual frameworks will help you to identify key data that needs to be collected and analyzed, to organize the data, and to understand the linkages between different types of data. (See the Annex for some examples of tools and frameworks.)

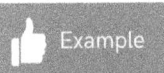

> Mapping tools that allow the user to visualize the system elements, relationships, and outstanding questions help to simplify a real-world situation so that linkages between different parts of the system can be more easily understood. They can help to provide a 'roadmap' to bring the user from where things are at the moment to where they could be. See the Annex for a list of tools.

❹ Coordination in assessments

Very often, multiple agencies may be able to use the same information. Coordination with the public sector, UN systems, and interagency groups can help agencies either work together on areas of mutual interest or focus their assessment efforts on sectors or geographical areas not covered by other agencies. Using tools such as standardized interview forms or report templates can facilitate joint analysis across agencies or within a cluster. This is especially important when assessments identify needs that fall outside the mandate of your organization (see also *Core Standard 2*).

5 Scope of work

Assessment scopes of work should take into consideration: 1) the technical and geographic mandate of the organization or project undertaking the assessment, 2) knowledge of the technical and geographic scope of existing or assessments or those planned by other agencies or organizations, 3) an understanding of the informational needs of decision-makers, and 4) resource availability. The assessment should consider all groups that may be affected by the crisis. The scope should be clearly communicated to community members in order to avoid creating expectations.

6 Mapping affected households, enterprises, and market systems

In developing the assessment scope, the approach should be dynamic, considering how affected markets, households, and enterprises operated before the crisis; how they were impacted by the crisis; how they cope now; and what potential they will have in the future. At the household level, livelihood strategies depend on effectively integrating assets and skills, social and economic relationships, and access to both consumption and output markets. Households may have several, diverse sources of income and multiple contributors to household income. It is important to understand the balance and trade-offs between income sources in relation to conflict, the environment, and gender dynamics.

Similarly, the success of enterprises in weathering the crisis period depends on a set of internal factors, such as human and technical capacity and capital, and on their interactions within larger market systems, including other enterprises, customers, financial and non-financial services, inputs, commodities, infrastructures, and regulatory frameworks. Assessments should recognize this complexity and identify and analyze the interdependencies involved.

Economic strategies for households and enterprises should always be embedded within the wider economic, political, and institutional contexts. The assessment team should attempt to include these contexts in their assessment or bring in outside expertise on the political and cultural contexts. This is particularly important in conflict environments, where programming needs to be aware of the dynamics of the conflict, including the roles of different actors, and how they relate to targeted markets.

Other considerations for the scope are:

- assets, skills, and capacity of market institutions and supporting structures
- key social and economic relationships and power dynamics, including gender
- governance within targeted industries
- linkages and accessibility to government schemes and programs
- relevant policies and regulatory frameworks for key industries and economic activities, including, as relevant, financial services, IDPs and refugees, labor, consumer protection, tenure and property rights
- availability, access, and status of key market infrastructure and formal and informal financial services
- related issues in natural resource management and conservation
- the impact of the crisis on markets, households, and enterprises, including how they operated before the crisis and how they respond on an ongoing basis following the crisis.

See also the *MiSMA*.

7 Initial and rapid assessments

Initial assessments and/or rapid assessments are quick methods to gather basic facts about how a disaster has affected market dynamics. Initial and rapid assessments should inform immediate response priorities and highlight areas for further investigation (see the Annex for some examples of tools and frameworks). It is important to keep in mind that these rapid assessments may be time-bound. The market context may change quickly. Consider economic and environmental factors in determining the timing of the assessment, and determine when updates to the assessment may be required based either on time passed or milestones reached in the recovery period.

Example

In Niger, community focal points alerted the response team to increasing numbers of people moving towards Lake Chad. The team launched a rapid assessment that revealed over 10,000 newly arrived people living on islands in the lake, with massive unmet needs for health care and clean water.

8 Sources of information

Assessments should first review existing research and information on livelihoods and economic activities prior to the conflict. They should rely on local sources and local actors, such as heads of households, storekeepers, and traders, as well as macroeconomic, political, and international sources. The information methods should be sensitive enough to identify hidden sources of information, such as marginalized groups and informal or black-market economic activities. Assessments should triangulate data through multiple sources, including assessments from other organizations, and, where possible, use both primary and secondary data sources. In high-risk situations or in rapid-onset hazards, programs may not be able to engage in a fully collaborative process or arrange collection of primary data.

9 Unnecessary data

Do not collect unnecessary data. In many cases the same data is already being collected by others. Collect primary data only when you have a clear understanding of its potential use. Primary data collection can also give a voice to key stakeholders affected by the problem.

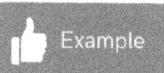

Example

The Vulnerability Assessment and Mapping (VAM) unit of the World Food Programme (WFP) has been gathering market price information on the most commonly consumed staples to strengthen food security analysis. This data is now consolidated and available online as the 'VAM Food and Commodity Prices Data Store'. The WFP's database benefits from a long history of price data collection, compiled by WFP country offices or collated from national government agencies and partner organizations. It provides a source of information and analysis to aid professionals, academics, students, and anyone interested in food security, and specifically in staple food price dynamics. It is often not necessary to collect pricing data as this resource provides good information on price trends over a long period of time.

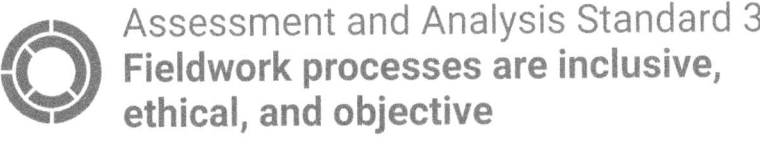

Assessment and Analysis Standard 3
Fieldwork processes are inclusive, ethical, and objective

Assessments gather data from a wide range of stakeholders using ethical, objective, transparent and inclusive methods, with special attention to vulnerability and coping mechanisms. Participatory methods are preferred when possible.

Key Actions

- Train enumerators properly, to improve data quality. Timing or seasonality, security, transparency, and potential biases should be addressed as part of the training.
- Gather assessment data using objective methods to ensure data quality.
  - Tailor methods according to whether you have physical access (e.g. mobile data collection).
  - Pilot the data collection tools and adjust as necessary.
  - Cross-check and clean data while in the field.
  - Consider use of IT for real-time monitoring. It allows for checking data immediately to see if something is wrong.
- Gather assessment data using inclusive, participatory, and ethical methods which do not compromise participants' security and mental well-being. (Note that inclusivity and participatory methods may sometimes be in conflict with security, so teams will need to use good judgement.)
  - Inform participants of the rationale for assessment and data use (informed consent).
  - Give participants anonymity (unless advised otherwise).
  - Ensure that the data collection process does not put participants in harm's way or waste their time.
- Put systems in place to improve data quality in the field.
- Ensure that assessment team composition reflects the gender, age, disability, and cultural variations of the communities participating in the assessment.

- Aim to engage the full range of actors who may be involved in economic activity (e.g. wholesalers, retailers, processors, buyers, and others along the value chain).
- Conduct interviews in a language and phrasing familiar to the participant.
- Clarify expectations ahead of the immediate interview process (avoid making any promises).
- Respect local customs during the interview process.
- Schedule interviews, observations, or site visits at different times of day to ensure that all daily economic activity or potentially hidden populations are captured during the assessment.
  - Schedule interviews/observations to monitor weekly or key market events.
  - Consider and mitigate potential assessment fatigue on households who have been repeatedly assessed and are traumatized (see also *Sphere Handbook Protection Principle 4: Protection of Human Dignity*).

Key Indicators

- The sources of data are varied, accurate, and high quality, and the information-gathering process is collaborative and participatory when possible.
- The methods used to collect data are sensitive to the biases of informants and interest groups, while factoring in the potential for intensifying conflict.
- Data are cross-checked with multiple sources and methods.
- The methods used do not put at risk the security of those conducting the survey or those surveyed.

Guidance Notes

1 Security of assessors and informants

The location or timing of an assessment interview may constitute a risk to either the assessment team and/or those being surveyed. The assessment team should consider the security of interview locations in determining appropriate places and times to conduct interviews. At no point should an assessment interview put either the interviewer or the interviewee at undue risk of physical or psychological harm (e.g. reliving trauma). Informed consent should be obtained where possible.

2 Biases and interest groups

The assessment should be sensitive to different interest groups and to possible bias among informants. Researchers should use non-leading questions and validate with multiple sources to determine if data is accurate. Cross-checking, triangulating, and using multiple methods is encouraged.

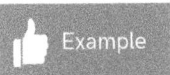
Example

> Program staff wanted to understand farming practices in order to provide appropriate recovery packages. When the enumerators asked, 'Who makes decisions on planting?', nearly all respondents said it was the husband in a household. Eventually one of the female enumerators explained to the team leader that women would be shaming their husbands if they admitted they made decisions on planting. The question was reworded to ask who was doing work on which crops, to better capture the differences of men and women's agriculture work preferences.

3 Transparency

Clarification of expectations is essential: be clear on how the data will be used, and what will (or will not) be given in return for participation in the assessment.

④ Choice of methods

Choose methods that are simple, concise, and capable of generating gender and age-specific information, and, if possible, wealth and livelihood disaggregated information. Apply methods that, while culturally sensitive, encourage and facilitate the participation of marginalized groups (including women, people with disabilities, and the elderly), including those who do not have the right to speak publicly. (See tools in the Annex, particularly the CLARA tool, for more details on methods generally and on including marginalized groups.)

⑤ Timing

Be aware of periods of increased vulnerability, including seasonal variations in supply/demand, price, and the hungry season.

> The baseline for a project was done during a particularly good harvest year. When a drought took place the following year, the organization responding needed to use additional data sources to understand what could be considered a 'normal' or average year in order to provide the appropriate levels of assistance.

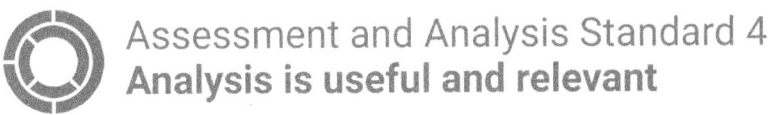

Assessment and Analysis Standard 4
Analysis is useful and relevant

Analysis of data and information is timely, transparent, inclusive, participatory, objective, and relevant for programming decisions.

Key Actions

- Completely clean and organize data before analysis.
- Coordinate the analysis with local humanitarian actors (public and private).
- Challenge your assumptions once primary data comes in. Triangulate as needed.
- Consider trade-offs between in-depth analysis and the need to share results rapidly for decision-making and programmatic response.
- Ensure the plan for the analysis is adequately captured in the preparedness, planning, and fieldwork stages (see also *Assessment and Analysis Standards 1, 2, and 3*) and in the assessment plan.
- Review the format of the analysis and presentation of the results to ensure they can be understood by a range of users (that is, not just specialists).
- Validate the results of the analysis with a diverse group of community members and other stakeholders.

Key Indicators

- Analytical outputs use state-of-the-art methodologies and tools when possible (such as remote sensing, big data, mobile data collection, and crowdsourcing, rather than econometrics and statistics), but make sure complex analyses return clear and simple findings.
- Analytical results are produced in a timely manner and provide useful and detailed information (e.g. by target group, location, and type of activity) for program design.
- Program model and design are grounded in solid analytical findings.
- Stakeholders at all levels are involved in validating the analysis.
- Data is treated as a public good and shared to the greatest extent possible considering aspects of accountability, security, and protection.
- Monitoring elements are teased out from the analysis.

Guidance Notes

1) Speed vs. depth

The trade-off between in-depth analysis and the need to quickly share results is especially important during an immediate response phase, but is present in almost all assessments. One of the keys to getting the balance right is knowing who will use the information, what level of detail is required for decision-making, and how these questions relate to the overall scope of the assessment. If some of the analysis is likely to take longer, the team can also consider multiple reports; get simple data out quickly where possible and later release analysis reports that require more time.

2) Review scope and consider the audience

In beginning your analysis, start with a clear understanding of the scope. What is the main intent of the whole exercise? Based on the plan that you developed earlier, which key questions do you want to answer? Think also of the final users of the analysis. What information do they need to make good program choices? Will they respond better to numbers or qualitative representations of the data? Understanding the preferences and uses of the completed analysis allows the team to focus on the most important elements of the analysis and not waste time doing calculations that are unlikely to be used.

3) Validation

Once the analysis is nearly finalized it should be validated with diverse stakeholders at all levels, to help ensure that the analysis is useful and relevant in their contexts. Validation most commonly takes place in the form of a meeting or workshop where the results are presented, and those who have knowledge of the situation provide information on whether the findings seem reasonable or 'ring true'. Ensure that the voices of vulnerable groups (e.g. women, the elderly, people with disabilities) are heard during the validation process.

④ Managing bias

Be upfront about any possible biases that are inherent in the analysis and its presentation. By remaining aware of and focused on potential sources of bias, staff can enable the truest respondent perspectives to come out. The use of technology can sometimes inadvertently introduce new biases if it reduces transparency, requires a certain level of education, or fits into prescribed gender roles. Try to use presentation approaches that make the information available in multiple ways and to a wide variety of stakeholders.

⑤ Stay flexible

Be prepared for surprises and be open to moving in directions that are different to those you had expected. Assessments do not (and should not!) always confirm what we think we know.

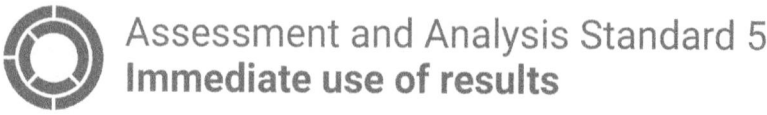# Assessment and Analysis Standard 5
Immediate use of results

Immediate steps are taken to ensure that assessment results are shared and used in programming, policy, advocacy, and communication decisions.

Key Actions

- Ensure that assessment data and analysis are quickly compiled into reports, summaries, and presentations.
- Conduct response analyses to determine the appropriate response options based on assessment findings.
- Use assessment results to inform program design or to modify existing programs.
- Disseminate results to relevant internal and external audiences and make them widely available in the public domain.
- Develop a format for presenting findings, conclusions, or results that is the most relevant and accessible to the audience.
- Translate assessment results into relevant language(s) to expand their audience, availability, and comprehension. Share assessment results with populations consulted.
- Use assessment results as advocacy tools, where appropriate, to influence policy decisions.

Key Indicators

- All economic recovery projects are informed by an assessment.
- Assessment data and analysis informs the program model and design.
- Assessment results are circulated to provide appropriate guidance to decision-makers (see also *Sphere Handbook Core Standard 2: Co-ordination and Collaboration*).
- Assessment results are published and promptly made available to relevant stakeholders in order to maximize their influence on decision-making processes, taking into consideration sensitivities that may exist.
- Assessment results are communicated in a language and in a format that is clear and appropriate to the respective audiences.

Guidance Notes

1 Response analysis

The response analysis includes listing a range of potential response options (including cash-based, in-kind procurement options, or market support options); analyzing feasibility, gaps, and priorities; exploring market risks and opportunities; and, finally, prioritizing the most relevant responses. The analysis should be as participatory as possible, allowing for consensus-building and communication with stakeholders. When prioritizing responses, some criteria to consider include cost efficiency and effectiveness, market impact, institutional capacity, individual preferences, and political acceptability.

2 Dissemination

Distributing the results of assessments encourages cooperation and collaboration in economic recovery programming. Engaging decision-makers and working with partners and local authorities throughout the assessment process brings assessment results to everyone's attention as soon as they are generated, and encourages trust and cooperation. When an assessment indicates that action is required, a joint presentation to the key decision-makers (donors, NGOs, government) can build momentum and a sense of joint responsibility and ownership (see also *Core Standard 2*). The results of the assessments and the decisions taken should be clearly communicated to the affected populations and market actors.

Example

The International Rescue Committee (IRC) leadership team in Niger has maintained regular formal and informal channels of communication with UN agencies, donors, and local authorities. Regular updates keep donors informed of the evolving situation, which in turn makes it easier for the IRC to renegotiate projects and contracts as well as secure funding for new projects to meet emerging needs. In a few instances, the information provided by IRC has prompted a response by another international NGO. However, there have been challenges with follow-through on commitments to launch activities that are made following the distribution of an IRC multisector assessment by OCHA. A functioning official coordination system was missing for some time, which made it difficult to hold actors accountable.

③ Appropriate formats

The results of assessments should meet the specific needs and comprehension levels of different audiences. Decision-makers in a crisis often have hectic schedules and severe time constraints. The type and length of the informational product and level of technical detail should cater to the requirements of the user. For example, senior program managers may require a one or two-page brief, collaborating partners may want a presentation, and a longer technical report would be appropriate for monitoring and evaluation specialists. A meeting with community members or a producers' association would be appropriate for sharing assessment results with informants and affected people. Particularly large or comprehensive assessments and findings can be disseminated in more than one format. Maps and layouts can be very effective in representing information such as the distribution of physical assets, natural resources, people's location and movements, and other types of phenomena (e.g. impact of a disaster across a territory).

Assessment and Analysis Standard 6
M&E occurs throughout the program cycle

Program performance and impact is assessed throughout the program cycle, in an ongoing and iterative manner.

Key Actions

- During program design, develop a theory of change, logical framework or chain of expected results from any given intervention, and carefully and clearly articulate assumptions and risks throughout the program.
- Define appropriate indicators, at all levels of the logical framework, to measure each expected result throughout the length of any intervention.
- Determine appropriate data-collection methods and frequency for each indicator based on feasibility, timeframe for expected results, and how the data will be used.
- Design a strategy for analyzing monitoring data, disseminating results, and using data to revise and improve programs, including feedback to and from all stakeholders.
- Monitor programs through regular collection and analysis of data on key indicators.
- Monitor wider market changes that occur during program implementation.
- Carefully track associated program costs.
- Have regular, strategic follow-up meetings to review monitoring data, ensuring that programming on the ground is responding to the needs identified in assessments.
- Make strategic decisions about when and if an external impact evaluation is necessary or useful.
- Assess the relevance of initial assessments and analyses and whether they properly guided program decisions.
- As market conditions change, revisit and update original assessments and analyses with additional data or information on economic or livelihood conditions, and allow for programmatic change.

Key Indicators

- A theory of change, logical framework, and/or results chain is articulated during program design.
- Monitoring data is collected and analyzed on a regular basis.
- Programs use monitoring and evaluation data to test processes and key assumptions, check expected impacts, and make revisions as needed.
- Assessment and analysis are ongoing and integrated into program operations to allow for monitoring as the political environment and markets evolve.
- Monitoring data is further analyzed in an evaluation to determine whether a program successfully achieved the expected results (performance and impact) and whether the initial assessments were appropriate and useful for program design and implementation.

Guidance Notes

Logical frameworks or results chains

Program monitoring and evaluation should track project activities and outputs, and continually check the assumptions on which program activities and envisioned impacts are based – keeping a clear, documented link between program activities and desired impact. Articulating a clear theory of change, logical framework, and/or results chain during the program design stage allows for this. Indicators should relate directly to the expected sequence of results or changes, at all levels, in order to ensure appropriate data collection and analysis (see DCED Standards for Results Measurement in the Annex). Be open to unexpected results that are not in the logical framework.

❷ Regular monitoring

Ongoing monitoring, evaluation, accountability, and learning enables programs to adapt to the fluid, sometimes rapidly changing, contexts of markets in emergencies. Change in program targeting, design, location, or intervention can enhance program quality and is sometimes implemented through a phased approach (for example, different targeting in immediate response than during recovery), with regular communication with the donor aiding program quality and transparency. Rather than depending on annual or biannual evaluations, programs should collect and analyze monitoring data regularly to provide timely information for decision-making and program improvements. This is best achieved with regular and systematic monitoring that tracks program outputs, outcomes, and critical factors in the external environment. It should also be linked to successful outcomes and the expected impacts. This keeps the program responsive to changing conditions so that, if need be, either programs or the relevant indicators can be adjusted as needed. Lean data can be used to collect a small amount of information relatively frequently, taking advantage of relevant technologies as appropriate.

During the west African Ebola crisis, a technology platform gathered national data and made the full monitoring data available to all partners through an online dashboard. Those who accessed it or interacted with the analysis at workshops found it useful. For example, some partners redirected their mobilizers' activities in response to activity-level data on geographic coverage and mobilization methods.

❸ Ongoing assessment updates

Conflict and crisis environments are dynamic. Data collection and analysis need to be continuous in order to stay responsive to the rapidly changing environment. Revisiting assessments may identify unintended consequences of programming and can help organizations make adjustments to mitigate negative impacts on households or maximize any unexpected positive benefits of an activity (see also *MiSMA Standard 5*).

Enterprise and Market Systems Development Standards

- **Standard 1**
  Send market systems staff immediately after a crisis

- **Standard 2**
  Implement market system analyses early and adapt frequently

- **Standard 3**
  Be adaptive and risk aware

- **Standard 4**
  Work with existing market actors and use facilitation approaches

- **Standard 5**
  Support viability and growth of enterprises and market systems

3 Enterprise and Market Systems Development Standards

Enterprise and market systems development means supporting the economic and livelihood activities of individuals and businesses. This includes everything from self-employment to large commercial operations, whether formal or informal.

Livelihoods are the economic strategies that people use to hold, utilize, and transfer assets to produce income. Enterprises, including entrepreneurs and farmers, are the entry point for understanding the market system. Both livelihoods and market systems interventions may directly support households or businesses, or use indirect approaches to help an entire market system or value chain recover or function more effectively. These Standards encourage supporting livelihoods and market systems before, during, and/or after a crisis in order to support people's access to critical goods and services, in a manner that helps targeted households to stabilize or increase their incomes (see also *Core Standard 1*). Multistakeholder coordination (government, private sector, civil society) and cross-sectoral coordination are usually required (see also *Core Standard 2*).

These Standards are intended to guide programs that seek to stabilize or increase the income of target populations, and use of these Standards will minimize market distortion. For many businesses, this concept may be described in terms of increased sales, growth, and competitiveness; however, the end result will still be increased or more stable revenues.

These are examples of enterprise and market systems development:

- Supporting individuals with technical support or assets to start or restart businesses after a crisis in a viable market (e.g. IDPs restarting a business after a conflict).
- Facilitating commercial information flows in a market system, making information more accessible to a range of businesses (e.g. preferences on product quality, price levels, specific quantities, or types of goods desired).
- Introducing or strengthening the relationships between actors at different levels in the market (e.g. producers to buyers, transporters to wholesalers).
- Strengthening the relationship between actors on the same level of the market system in order to improve efficiency or quality (e.g. hosting a convention of wholesalers or importers in a given market).
- Encouraging others to facilitate or, on occasion, directly provide services that businesses need (e.g. specialized finance, veterinary services for pastoralists).

Enterprise and Market Systems Development Standard 1
Send market systems staff immediately after a crisis

Market systems technical staff are deployed with first responders to ensure relief activities are designed in ways that limit negative impacts on markets and livelihoods.

Regardless of the approach taken (market-aligned relief, market-sensitive relief etc.), the planning and design of interventions must ensure that market actors and existing networks are used to the greatest degree possible, long-term negative impacts are limited, and interventions are designed with maximum flexibility so they can respond as the situation changes.

Key Actions

- Market system technical staff participate in the development of assessments for other sectors to ensure input is gathered on the conditions of market systems, how responses may affect private-sector actors, and longer term recovery issues.
- Recognize that all relief affects markets and identify opportunities for relief to work with and through market actors in some capacity.
- Seek opportunities to support and use existing local private-sector suppliers and enterprises in relief operations, while considering equity and power dynamics.
- Collaborate and engage with local private-sector actors, both formal and informal, in assessment and program implementation. This may include chambers of commerce, cooperative networks, merchant associations, and small farmers' groups/federations.
- At the cluster level, ensure coordination with other organizations and government entities, and recognize that joint proposals should include or encourage sufficient flexibility to take into account existing market dynamics. (See also *Core Standard 2*.)
- Market staff work with procurement staff to avoid setting up a parallel system for goods and services delivery.

Key Indicators

- Staff members with markets expertise are deployed or identified in-country.
- The team explores opportunities for integrating market approaches into activities.
- Flows are sourced through existing local market actors when possible or else conditions and timing under which that will be possible are identified.
- Short-term solutions are designed to facilitate longer term goals.

Guidance Notes

❶ Longer term impact

Most post-crisis activities are focused on addressing immediate needs. Yet, emergencies also damage market functions and trade networks. Additional harm may be done if humanitarian interventions do not consider the potential of the local private sector to participate in recovery, as well as their recovery needs. Short-term activities that do not consider the wider market system are likely to undermine future recovery aspects and potentially destroy livelihoods. All project planning and design must consider the long-term impact of the activities.

❷ Working with procurement

The choice of who we procure from is critical. Explore the potential for strategic sourcing of goods and services that rebuild local markets and livelihoods (i.e. if in-kind is an appropriate response, consider buying from local private-sector actors, despite higher costs, where feasible. Donors, including USAID and the UK Department for International Development, often explicitly encourage this.) Ensure that procurements do not make existing power imbalances worse.

Enterprise and Market Systems Development Standard 2
Implement market system analyses early and adapt frequently

Market analyses occur alongside other needs assessments immediately after an emergency. Time is built in to monitor and update these analyses in a quickly changing context.

Key Actions

- When working in areas that are likely to face crises in the future, ensure pre-planning includes an assessment of the vulnerabilities for critical market systems (see also *Assessment and Analysis Standard 1*).
- Design emergency response activities from the beginning with an awareness of the risks, vulnerabilities, and opportunities for the livelihoods and markets where they are carried out (see also *MiSMA*).
- Consider the needs of specific groups (women, youth, people with disabilities, elderly, and disadvantaged ethnicities) in addition to the economic impacts when planning interventions. Recognize that the most marginalized will be the least likely to be captured in the data.
- Consider the existing relationships and knowledge of local market system actors and existing development program actors working in targeted market systems.
- Update assessments frequently, recognizing the changing nature of humanitarian contexts, and ensure ongoing assessments bring new and/or better information that builds on previous assessments and reflects the changing market system dynamics.
- Include information from the formal and informal markets when gathering information on demand/supply gaps.
- Consult with affected communities and market actors to find the best ways to share analysis findings, staying mindful of how specific populations (e.g. women, farmers, vulnerable groups, people with disabilities) and actors (e.g. traders, retailers) may be affected.

- Encourage market system assessments to be used when designing, implementing, and monitoring relief and economic recovery programming in other sectors; market staff should share them widely and support colleagues as required (see also *Assessment and Analysis Standard 5*).

Key Indicators

- Market vulnerability assessments are used to design activities that strengthen the resilience of the market system, and are used as a 'baseline' of the market system to help guide future relief actions, particularly in crisis-prone areas.
- The scope of the market assessments includes non-affected areas – wherever the supply chain extends – even when outside the areas most affected by the crisis.
- Market systems analyses include information on support services, government initiatives, private-sector groups, the formal policy environment, and informal social and cultural norms influencing the market system.
- Assessments and analyses are updated regularly to support changes that need to be made to program activities or verify that no changes are required and activities are working.
- Input from specific target groups (women, youth, farmers, people with disabilities) is understood and utilized in the response activities.
- Financial and market analyses are done in a manner that includes the communities and enterprises they affect, allowing those concerned to weigh the potential risks, returns, and mitigation strategies.
- Interventions include basic financial and market assessments of the enterprises, and show that enterprises selected for activities will be profitable and viable, in terms of both income and risk.
- Risk-mitigation strategies are incorporated into interventions to help enterprises prepare for shocks that could undermine their business.
- Vulnerable households are able to make informed decisions regarding specialization or diversification of their business activities.
- Relief is delivered to households through local market actors, leveraging pre-existing partnerships with private-sector partners where possible and appropriate.

Guidance Notes

1 Preparedness and pre-crisis planning

Organizations working in crisis-prone areas or areas affected by protracted crises should include vulnerability assessments in their market systems analysis. Assessing vulnerabilities and the scope of possible economic shocks prior to a crisis can improve preparedness and inform contingency planning efforts, which will increase the speed of future emergency responses, help reduce vulnerabilities, and increase resilience within the market system. Improving targeting and strengthening the market system prior to an emergency can reduce the negative impact of disasters on lives and livelihoods.

 Cautionary tale

In Odisha, India, two cyclones wiped out the maize crop in 2013 and 2015. Maize bulk buyers then stopped providing upfront credit because of the losses they suffered from farmers unable to repay. When credit for seeds and other inputs stopped, farmers stopped growing maize, which had the potential to increase food insecurity in the area.

2 Market systems analysis

Market systems will often be spread geographically across a country (or beyond its borders). It is important to understand how areas not affected by the crisis are also part of a critical market, so that they can be leveraged for a response. This will also help ensure that these market systems are not distorted by response activities. Actors in the wider market system may be able to contribute to the response and/or may be negatively affected by an emergency response if they are not included in the market analysis.

> **Cautionary tale**
>
> After an earthquake, humanitarian agencies distributed free agricultural inputs within the earthquake-affected areas. This distribution weakened the input supply market system by undermining the businesses of the farmers and traders from outside the affected areas, because their customers were receiving free inputs and no longer needed to purchase them. While a market analysis had been conducted directly in the earthquake zone, which identified the need for inputs in re-establishing livelihoods for those directly affected, the market analysis had not extended to the suppliers in the market chain and therefore the impact of these free hand-outs on the wider market system had not been considered.

❸ Minimum requirements for assessments

Early in the response, assessments may be quick and dirty, and this is acceptable because 'good enough' analysis is better than 'analysis paralysis'. However, each following assessment should add to and deepen the understanding of markets and livelihoods. Target households may be better reached by working through existing market actors (e.g. traders/vendors). Asset transfer activities (e.g. cash and vouchers) should work through and support local market systems where possible (see also the *Assessment and Analysis Standard* and *MiSMA*).

❹ Effect of relief boom

Forecasts of potential profits should take into account the possibility that a business may be vulnerable to a 'relief boom' – when enterprises and industries (such as restaurants or construction and transportation companies) are only viable while relief funds and workers are part of the economy. Caution should be used when assisting these types of enterprise and support given to enterprises to think post-boom. Looking at alternative strategies based on different possible outcomes might be helpful, especially in changing environments.

5 Involving stakeholders in analyses

Analysis should include the views and opinions of enterprises, communities, and individuals targeted for assistance, and be shared with coordination groups. Including multiple stakeholders in the analysis will help them understand the potential risks and returns in the short and long term, and assist them in making informed decisions. It will help ensure that the information and assumptions used in the analysis are reliable. Programs need to understand stakeholders' available resources, the performance of businesses to date, and the level of risk that businesses and households are comfortable taking on. When all stakeholders understand the analysis, they are in a much better position to make informed decisions regarding a specific approach. This reduces the risk of supporting activities that may further impoverish households (see also *Core Standard 2 and Core Standard 5*).

Example

The EMMA market assessment in Port-au-Prince, after the 2010 Haiti earthquake, looked at the market system for roofing materials. By speaking to stakeholders – small building-materials suppliers, large building-materials importers, the government, and humanitarian agencies – analysts could get a full picture of the needs and future demand for roofing materials. By talking with analysts and reviewing a market map, the people who worked in the roofing materials business were better able to understand how the role of others in the market system could impact their competitiveness. This helped them make more informed decisions about their businesses.

⑥ Informal social and cultural norms

Social and cultural norms play an important role in defining the incentives, relationships, and behavior of actors in market systems. Understanding the influence of social and cultural norms on issues of gender, caste, and tribe, and how these affect relationships within and between these groups, is particularly important in conflict and post-conflict situations where issues of identity are often drivers of conflict and social networks and trust have been disrupted and weakened.

Example

In Afghanistan many farmers were resistant to enterprise development programs that tried to remove middlemen from the market chain or change the location of product sales. The buying and selling relationships between farmers and sales agents had been set over generations and to change these would be breaking a social norm, as well as undermining trust within the communities.

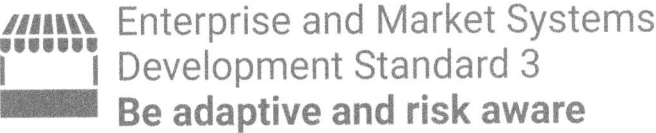# Enterprise and Market Systems Development Standard 3
Be adaptive and risk aware

Adaptive and flexible management techniques are used at all levels to ensure programs respond to the changing conditions of the market.

Key Actions

- Build in preparedness activities for pre-crisis programs working in areas with recurring or ongoing crises, and build in the flexibility to respond to unexpected crises through mechanisms such as 'crisis modifiers'.
- Design programs with maximum flexibility so activities can adapt to changing household and market needs as the context evolves. This includes initial design, budgets, management systems, and M&E.
- Ensure compliance, finance, procurement, contracts, and other operations teams are aware of the flexibility allowed in the grant, and help program staff find appropriate solutions to necessary programmatic changes.
- Establish monitoring systems able to capture data on changing market dynamics, particularly the availability and prices of key goods and services.
- Review monitoring data regularly and use it to make program changes in response to changing market dynamics.
- Hire staff with a collaborative and flexible mindset, and put systems in place that allow for adaptive management.
- Conduct frequent activity and program reviews and adjust program activities according to the findings. Create systems for stakeholder feedback and incorporate the data into the program reviews.

Key Indicators

- Emergency and early recovery programs are designed to be flexible and to respond to changes in the market system or enabling environment. Programs use adaptive management techniques.
- Development programs are designed to be flexible in the event of likely crises, and include crisis modifiers or other mechanisms to facilitate rapid response.
- Monitoring systems are adequately resourced and capable of gathering data on market dynamics in a timely manner.
- Staff have time and ability to regularly refer to and use market monitoring data to make program adjustments as necessary.
- Program activities are adjusted based on the most up-to-date assessments and analyses.

Guidance Notes

 Program flexibility

Donors and implementers have a joint responsibility for being responsive to market realities, and programs need to be designed with the ability to change approaches based on shifting market conditions. In development programs this can be done through the use of mechanisms such as a crisis modifier to transition to emergency response, while in relief and economic recovery programming it might be related to transitioning from in-kind distribution to cash-based programming based on changing market contexts.

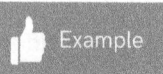

> The Pastoralist Areas Resilience Improvement through Market Expansion (PRIME) project in Ethiopia began just a few years after the drought of 2010-11. As a result, the development program had flexibility written into it, in the form of a crisis modifier, to allow funds to be used for early response and emergency activities to help 'protect development gains'. The redirected funds were used for fodder provision and livestock offtake when rains failed a second time.

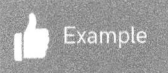 Example

The International Rescue Committee (IRC) has responded to humanitarian needs in southeast Niger since 2013. Facing a shifting context and challenging operational environment, the team developed a 'rapid response' mechanism, funded by multiple donors. With dedicated funding and staff of this mechanism, information received from the focal points and committees network can now trigger multisector assessments using household surveys, key informant interviews, and focus group discussions. A scoring system flags critical issues and prioritizes hard-to-reach and highly vulnerable areas. These assessments have led to supply of rapid food, non-food items, and water, sanitation, and hygiene activities, as needed.

2 Resourcing monitoring systems

The ability to effectively adapt to changing circumstances depends on the quality and timeliness of the information provided to program staff and management. There are numerous cases where assessments are delayed and results are late or not comprehensive enough to be useful for decision-making. Where possible, programs should leverage existing mechanisms, such as the Famine Early Warning Systems Network (FEWSNET), World Food Programme (WFP), and others, that capture prices and other relevant information which can provide insights into changing market dynamics. Monitoring systems therefore need to focus on timely collection and analysis of data, and need to be adequately resourced to carry out this important function (see also *Assessment and Analysis Standards 4* and *5*).

③ Adaptive management

Especially in highly volatile contexts, program managers and team leaders should assume that data-driven program adjustments will need to be made. Programs should build in frequent assessments, use indicators that allow for market changes, be able to reallocate resources according to identified opportunities and constraints, and have systems in place that track activity changes. Project staff should ideally have multiple complementary skill sets (e.g. cash, markets) and should be hired specifically for their creative thinking and collaborative problem-solving skills. Leadership should create space for discussion and support those who seek new solutions or ways of working.

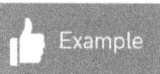

The program leadership and donor jointly agree at the beginning of the project on the impacts and outcomes that are most important. Activities, and some indicators, may be changed as long as the new activities contribute to the original impact priorities. These changes will normally be based on field assessments and other market data and will be documented in project reports. The budget is designed with a degree of flexibility to allow for activity-level changes.

Enterprise and Market Systems Development Standard 4
Work with existing market actors and use facilitation approaches

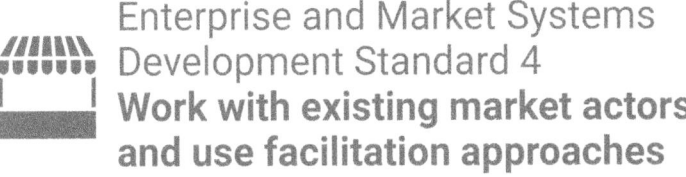

Interventions work through existing structures and are designed and implemented with long-term sustainability and resilience in mind.

Key Actions

- Build sustainability and resilience by ensuring local ownership of activities and by implementing, whenever possible, through partners already active in the market system.
- Establish impartial and transparent communication with formal and informal market actors while designing activities.
- Take a facilitation approach to implementation and maintain separation between project activities and market system activities. Implementers should not take on market functions.
- Work at multiple points in a market system to address the root causes of constraints and vulnerabilities.
- Design activities so that the exit strategy is a natural and clear transition.
- In crisis-prone areas, include activities that address market system vulnerabilities in all humanitarian and recovery programs. Ensure that clusters or coordination groups are aware of these vulnerabilities and the actions being taken to address them.

Key Indicators

- Programs engage with the public and private sector as partners and avoid directly entering the market. When they must do so, they ensure that an exit strategy is in place from the outset.
- Programs are clear from the beginning about their reasons for working with enterprises and clearly state how working with enterprises will benefit the target population as consumers, suppliers, or employees (for employees, see also *Employment Standard 1*).
- Subsidies are time-bound, informed by analysis, and used selectively to stimulate a market response.
- Short-term solutions are designed to facilitate longer term goals.

Guidance Notes

1 Systems-level support

Programs should consider working at several points in the market system or value chain to improve the effectiveness of interventions. Activities implemented with processors, wholesalers, or government (indirect or systems-level support) may provide as much benefit to the targeted population as direct interventions targeting individuals or households (see also *Core Standard 5*). Use a response analysis tool to analyze the potential outcomes of indirect and direct methods, and select activities based on the desired results (scale, behavior change, impact on specific target groups etc.). Results chains can also be helpful for determining whether to use direct or indirect methods.

Program staff sometimes fail to understand that intermediaries (e.g. wholesalers and middlemen) can play an important role in the market. As part of any assessment, determine the roles played by intermediaries in the market system and how they can be leverage points for facilitation activities.

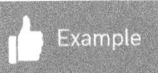
Example

In order to support poor farmers in Azerbaijan, IRC repaired the electrical lines to a nearby grain mill, significantly reducing the distance (and therefore transport cost) for farmers needing to mill their wheat. As a result, their profits increased. The by-product of the milling was made into animal feed, thereby improving the nutritional state of local livestock, as inexpensive feed was now available.

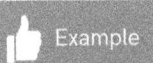
Example

In Haiti, costs for basic food items soared in areas where small wholesalers had no place to store bulk goods. By supporting the repair of small storage depots, organizations helped reduce the costs of staple foods for poor consumers. Wholesalers could again purchase in bulk and sell at better rates, thereby improving food security in the area (more people were able to afford their goods). As a result of selling more, wholesaler profits went up.

❷ Selecting private-sector partners

When using a market facilitation approach, the selection of appropriate private-sector partners is important and is likely to affect outcomes. Things to consider in selecting partners are:

- Leverage:
  - Is the partner at the appropriate point in the market chain to help make the change that the project would like to see?
  - Does the partner have enough influence with peers and customers to help make the change that the project would like to see?
  - Does the partner have a relationship with the government that will positively or negatively impact the project outcomes?
- Will:
  - Has the partner chosen freely to participate?
  - Does the partner realistically have time to participate given business needs, family needs, and recovery activities?
  - How much will the partner invest directly in the activity? (This is a proxy for commitment.)

- Skills/capacity:
  - Does the partner have the technical skills to be engaged in the activity? If not, do you have the time to build these skills?
  - Does the partner have the business skills to be engaged in the activity? If not, do you have the time to build these skills?
  - Does the partner have the financial capacity to achieve the desired outcomes?
  - Does the partner have staff with the leadership skills to drive the activity independently? Will they be available to the project over the project life?
  - Will the partner be able to continue the work after the project ends?

Engagement with the private sector in a participatory manner during the design of activities is critical to building a sense of ownership within the business community and the targeted enterprises. It offers the opportunity to bring in additional expertise and resources. Existing enterprises – whether buyers, processors, or producers (farmers) – can provide the sustainable leadership needed to drive innovation.

Market opportunities and constraints generally require a coordinated response by enterprises in an industry or subsector, and this takes trust and a willingness to collaborate. Consider collaboration with existing industry structures, such as chambers of commerce or trade associations, even for humanitarian responses, as they usually have well-formed networks. Programming should have the means to strengthen collaboration and relationships between market actors and among targeted enterprises and individuals (see also *Core Standard 2* and *5*).

❸ Partnering with the public sector

Enterprise and market system development need to acknowledge the roles of both the private and the public sector. It is rarely a good idea for the public sector to take the lead on business activities; however, in many settings, various government agencies at local levels may offer services for enterprises, whether formal or informal. When appropriate, programs should engage with these agencies rather than setting up parallel service delivery systems, and should try to align activities with strategic governmental priorities.

④ Subsidies

Often in the form of grants or matching grants, subsidies can be effective in helping to replace assets (such as equipment) that have been destroyed during a crisis (see also *Asset Distribution Standards*). Subsidies can also be used to expand and explore new markets and for innovation, such as demonstrating the potential of an improved technology. However, subsidies are unsustainable and should be time-bound and used selectively. Over the long term, subsidies tend to distort the market and should be avoided when possible. If absolutely necessary, organizations should plan from the beginning how long the subsidies will be in place for and to communicate this clearly to recipients and other stakeholders.

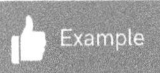

In Ethiopia, farmers were offered a 30 per cent discount voucher for solar pumps that were purchased at a trade fair or within 10 days of the trade fair. The discount encouraged farmers to try new technology, but they understood that the discount was for a limited time only. They became aware of the new technology through the trade fair and the distributor could give them information on his other products, such as solar lanterns, and the location of his shop (for potential future purchases). The project covered only the cost of the discount so the distributor could maintain his margin and continue expanding his distribution channels, making the new technology available to areas further from the town center.

⑤ Transparency

It is important for relief, recovery, and development programs to transparently engage with market actors from the outset of their activities. Market actors should be consulted and made aware of the program's approaches and objectives for working with enterprises, what the key activities are going to be, and how enterprises will be engaged. Failure to engage market actors in the design of relief, recovery, and development activities can result in lack of buy-in from market actors in the activities themselves.

> **⚠ Cautionary tale**
>
> In the Philippines, the Chamber of Commerce was not consulted during the design of the initial response. In the subsequent recovery activities, members of the chamber were skeptical of the commitment of NGOs to work with the private sector because of the experience of having their businesses disrupted by the initial response.

Enterprise and Market Systems Development Standard 5
Support viability and growth of enterprises and market systems

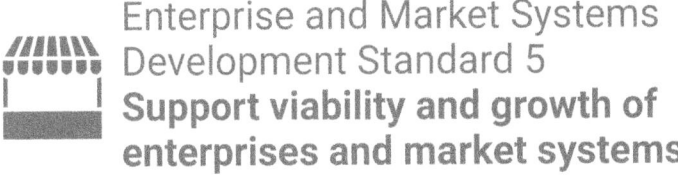

Activities, including emergency response, are designed with resilience and sustainability in mind, focusing on the stabilization and/or growth of livelihoods and enterprises of all sizes.

Key Actions

- Use available tools to design activities that give each stakeholder an appropriate role, and ensure that long-term owners of an action or market function are taking on that role as soon as possible. (See the Annex for resources and frameworks.)
- Incorporate risk mitigation strategies into interventions to assist enterprises in preparing for potential shocks and to strengthen resilience within the entire market system.
- Consider labor issues when evaluating enterprises and farms, and the potential need for skills upgrading, especially for target groups (see also *Employment Standard 2*).
- Include supporting functions and economic infrastructure (e.g. financial services, transport, warehousing) in market system assessments, and determine how emergency responses can contribute to (or avoid creating bottlenecks for) those functions.
- Make sure enterprises and farmers have access to reliable market information that allows them to change prices, know what products to produce, and where products can be sold.
- Strengthen mutually beneficial relationships between enterprises. Examine and reinforce horizontal (business to other businesses on same level) and vertical (business to businesses above or below them in the market chain) links between enterprises.
- Help enterprises be aware of relevant legal and regulatory issues, especially those identified during market systems analyses.

Key Indicators

- Risk mitigation strategies are incorporated into interventions and plans.
- Program activities stabilize or increase the income of targeted enterprises and households.
- Enterprises have affordable access to supporting services.
- Programs work at multiple levels in the market system.
- Enterprises and farms have access to reliable market information.
- Enterprises are aware of relevant regulations and can abide by them to the greatest extent possible. Programs also work to address enabling environment constraints related to formal and informal rules, laws, and regulations.
- Enterprises are encouraged (and given training if required) to operate in a socially and environmentally responsible manner.

Guidance Notes

 Risk mitigation strategies for enterprises

Support market actors to plan for shocks, including disaster contingency planning for the most common or greatest sources of risk in their context. This may include financial and organizational planning. Increasing the capacity of market actors to anticipate risk, and determine their exposure and risk tolerance levels, improves investment decisions, cost-benefit analysis, preparedness, and their own and market system resilience.

> An agribusiness program links smallholder farmers to a potato-chip processor. The smallholder farmers opt to move away from crop rotation to specialize in the potato variety requested by the processor, with the promise of handsome returns for their crops. Two years later, a potato blight wipes out their potato crops – and therefore their profits for the year – leaving them with no alternative crops to rely on for income or means to feed their families. The implementing agency should have provided better information to the farmers on the risks of specializing in one potato variety, and assisted them to look at how to mitigate the risk. For example, they could have grown potatoes as well as another one or two high-value crops, rather than focusing on one high-value crop.

❷ Increasing income

All economic recovery programs should directly or indirectly increase incomes, or at least make them more stable. Any program that does not do this is likely to be designed as pure aid, rather than enterprise development or livelihood strengthening, and should be described as such. For many businesses, especially small and micro enterprises, this concept may be described in terms of growth and competitiveness. However, the result will still be improved and more stable revenues. The success of enterprises and communities depends on reliable and diversified income sources (see also *Core Standard 1*).

❸ Linking enterprises to business service providers

Enterprises are part of larger market systems and require access to a range of products and services to succeed. Transport, finance, storage, and repair services are examples of business services that an enterprise may require to be successful. Assessments should include information on services that exist and how they are utilized; network analysis can be a helpful tool for doing this. If essential services are not available and/or are not accessible and affordable, then it may not be viable to support the enterprise. Special consideration should be given to linking enterprises to financial services (see also *Financial Services Standards*).

4 Training

Programs should avoid giving trainees payments (sometimes called 'sitting fees') to attend training or workshops as it distorts the incentives for participation. While it is understood that training takes time from productive activities and household responsibilities, it is generally recognized that if the training is considered valuable by the participants, then it is worth their time to participate. If it is not considered valuable (people do not want to attend without payment), then the project should consider whether the topics are truly relevant to the participants and whether their relevance has been properly communicated. Other costs incurred by participants (such as transport, lunches, lodging) are distinct from sitting fees and can be reimbursed or handled separately. There is no restriction on asking trainees to pay to participate because willingness to pay also shows that they value the topics presented.

5 Market information for enterprises and farms

Programs should ensure that targeted enterprises and individuals have access to regular and reliable market information. This includes knowing who is buying the goods or services in question, what quantity and quality specifications they prefer, what price different buyers are willing to pay, where to go for needed supplies or services, how prices change seasonally, what administrative and fiscal regulations must be abided by, and other information. This need for market information is as true for home-based businesses and farms as it is for large companies. However, for many reasons (remote areas, lack of access to communication technology, misunderstandings or outdated information, or competition), this market information does not flow through the market system as it should. Without it, an enterprise risks making uninformed business decisions and, potentially, losing money. Indeed, enterprise development programs sometimes focus exclusively on improving the flow of market information, with considerable success.

In northern Uganda, village-level agri-input shops were linked directly with certified seed suppliers in the capital. (For many years, farmers had gotten their seed supply through government distributions.) This provided the shops with higher quality seeds in a more timely manner, but also allowed the suppliers to learn that changing their packing to smaller 1 kg and 2 kg sizes would make them more affordable for farmers. This increased their sales and their reputation for good-quality seeds because larger bags did not need to be opened (making them vulnerable to contamination and switching) in order to sell seed in small quantities.

6 Regulatory framework for enterprises

Organizations committed to developing enterprises should be knowledgeable about the regulatory framework for the market(s) in which they operate and ensure – to the greatest extent possible – that the enterprises they work with are legal and compliant. (Note that an enterprise can be legal but informal.) This includes compliance with required licenses, taxes, and other regulations. If the enabling environment tolerates widespread informality, then programs may work with informal enterprises but should, where possible, support an advocacy and policy agenda that advances more formal infrastructure.

After the Indian Ocean tsunami in 2004, a few humanitarian organizations began to provide mechanized fishing boats to affected fishing communities instead of replacing the traditional fishing catamarans that had been lost during the tsunami. There was no analysis of the customary fishing mechanisms of the community. The Marine Fisheries Regulation Act of 1994 forbade mechanized vessels from fishing within eight nautical miles of the coast, yet these mechanized boats were used to fish within these parameters, destroying traditional boats and the ecology. Once their attention was brought to the issue, humanitarian organizations began providing traditional catamarans. A few organizations provided these boats through a share partnership mechanism which, in turn, helped reach more members of the targeted population who had limited available resources.

In informal markets, there are often informal requirements for operation that are controlled and enforced through social networks and links to those who have power in the market. How they affect the viability of targeted businesses and livelihoods needs to be understood and accounted for in programming. Business-related regulations and procedures can create negative incentives and discourage the creation and operation of businesses. Due to the institutional weakness in post-conflict settings and the reduced ability to enforce laws, applying the rules may become even more challenging, confused, and poorly transparent. These conditions represent an obstacle to the recovery of business and must be gradually removed by introducing appropriate incentives and creating adequate institutional capacities. While the issue is to be tackled at the central governmental level, specific measures could be decentralized, provided there are local capacities to apply them.

7 Corruption

Corruption is a common feature in post-crisis settings because the existing conditions allow it to spread. In turn, corruption reduces the credibility and popular support of the government, and, in turn, fuels political instability and conflict. When corruption and abuse of public power (e.g. illicit taxes) are strongly and negatively perceived by the population, lawful businesses struggle to start and grow. Corruption and abuse should be tackled to support the growth of the economy and contribute to longer term stability. In post-conflict settings, aid flows can also be perceived as an opportunity for further illicit businesses if they are misspent or wasted. Good accountability and risk mitigation mechanisms should be put in place.

8 Responsible enterprises

Investments in 'responsible' businesses contribute to peace and stability. Such investments should be transparent, compatible with local values and development needs, and conflict-sensitive. Moreover, responsible enterprises do not deplete natural resources but instead actively contribute to their protection and renewal. A number of codes of conduct for enterprises have been developed to promote economic growth, human rights, environment protection, and social development. Some of them are sector-specific (e.g. oil, gas, minerals, metals) and others are related to labor standards (see also *Employment Standard 1*).

Asset Distribution Standards

Standard 1
Asset programming responds to identified needs

Standard 2
Asset programming stimulates recovery without undermining local markets

Standard 3
Productive assets are protected

Standard 4
Asset replacement is fair and transparent

Standard 5
Assets expand and diversify livelihoods

4 Asset Distribution Standards

For the purposes of this Standard, 'assets' are specifically *productive assets*: resources used to generate or facilitate income, profit, and/or value. The terms 'assets' and 'productive assets' are used interchangeably in this chapter. Other livelihood assets, such as social capital, human capital, financial capital, and natural resources, are also important for economic growth, but are not fully articulated here. People can make use of physical assets in two ways: 1) they can own or directly control them, or 2) they can have access to resources that do not belong to them.

After a crisis, asset programs must respond to individual and household needs without undermining local markets and address issues of transparency, equity, and longer term impact. Asset distribution programming safeguards and/or restores assets that are necessary to productive livelihoods through three approaches:

1. *Protecting existing assets* from being consumed, lost, or sold to meet immediate basic needs.
2. *Replacing assets* lost as a result of the crisis, by evaluating the value of assets lost and replacing them accordingly, or disbursing one type of asset, such as tools, to all target households. New or adapted assets may be provided when former livelihoods are no longer feasible.
3. *Facilitating access to assets* for the expansion, adaptation, creation, or diversification of livelihoods after a crisis to either increase income or access new livelihoods opportunities.

While the first two approaches to asset programming focus on access and control of physical assets, the third approach may be linked with efforts to improve other types of capital (such as human, social, and natural capital).

There are many ways to achieve these aims. The most appropriate approach depends on the context, the goal of the program, and the target group. Common methodologies include:

1. *Cash transfers:* direct payments of money to an individual to enable them to purchase needed assets or avoid the sale of existing assets. Due to the efficiency and speed with which these programs can be implemented, the support they give to local markets and businesses, and recipients' decision-making power to choose what, when, and where to buy, cash transfers are often considered a particularly efficient and effective way to support program objectives. As an example: market sellers are given cash equal to the value of equipment lost in a market fire. Beneficiaries are free to use the cash for any business purpose, including starting a different enterprise than the one they had before the fire.
2. *Vouchers:* a coupon, piece of paper, or electronic equivalent that can be exchanged for goods or services. They function similarly to cash transfers but enable the organization to restrict the items purchased. The use of vouchers is a middle option that limits how the money may be used but may give recipients freedom to select between a few products or sources. An example of this: refugees are given a voucher booklet worth US$400 to buy household goods (blankets, curtains, pots, and pans) to replace those left behind. They can use any of 30 different shops in the city near their camp. They can choose what they need and are given up to six months to use the vouchers. Shop owners have agreements with the organization and receive payment for vouchers received monthly.
3. *Direct transfer:* Assets can also be purchased and directly transferred to beneficiaries, especially in situations where local markets are not functioning. Some donor regulations still make this the easiest form of asset distribution (administratively), though it is often less efficient than other

methods. For example, households that previously raised poultry are given chickens and feed to restart their livelihoods after a flood.

4. *Other:* In addition, other program approaches may be used for asset recovery and protection. For example, a financial access program might offer loan guarantees to small enterprises to help them access credit to replace assets lost in a crisis, or matching grants may be provided to encourage the expansion of livelihoods using new technology. Destocking is another type of asset protection activity that can be used to mitigate the impact of economic shocks.

Much of early recovery programming focuses on pastoral or agro-pastoral livelihoods and therefore livestock interventions are often a central part of asset distribution programming. The *LEGS* provide detailed guidance that is specific to protecting and rebuilding the livestock assets of crisis-affected communities.

Websites and resources are available to provide deeper explanation and analysis of these methodologies. Several sources are listed in the Annex.

For programming in asset protection, asset replacement, and assets for livelihood expansion, diversification, and adaptation, Asset Distribution Standards 1 and 2 apply. These are cross-cutting and serve as a minimum basis for all asset programming. Asset Distribution Standards 3, 4, and 5 specifically cover activities in the three approaches described earlier: protecting existing assets, reviving previous livelihoods, and developing new and diversified livelihoods. As you are designing and implementing programs in the areas covered in Standards 3-5, ensure you are responding to individual and household needs, as outlined in Standard 1, and stimulating recovery without undermining local markets, as indicated in Standard 2.

Asset Distribution Standard 1
Asset programming responds to identified needs

All asset programming, whether for protection, replacement, or livelihood expansion, should respond to beneficiary needs.

Key Actions

- Identify individual and household needs, capabilities, preferences, and aspirations.
- Ensure immediate basic needs are sufficiently met to avoid asset sell-off, immediate consumption, or sale of supporting items (such as sale of tools when seeds are provided).
- Ensure that vulnerable groups within the community (women, youth, the elderly, people with disabilities) are given special consideration when allocating assets.
- Make certain that all individuals (men, women, boys, girls) understand who is directly benefiting from the asset and why. Develop and publicize straightforward and transparent participant-selection criteria and program objectives.
- Mitigate for potential or ongoing risks (as a result of the program) to the physical security of individuals, their assets, and resulting income, and take steps to address these risks.

Key Indicators

- Assets provided or protected are critical to meeting individuals' and households' basic needs.
- Productive assets are identified, protected, replaced, and/or increased in an equitable manner.
- Individuals and households can articulate the goal of the asset program and why they were selected or not selected.
- Asset intervention has not contributed to community or household conflict or inequity.

Guidance Notes

① Needs assessment

To protect productive assets, it is necessary to understand local livelihood patterns and focus the intervention on those assets which have the greatest impact on household food and income sources. Agricultural households might produce a certain percentage of their yearly food consumption on their own, but if more of it is purchased, the immediate focus should be on maintaining access to food in the market, and on labor opportunities that provide that income. Intra-household dynamics must also be considered: ensure that assets are being used in a way that supports the whole household, that individuals are not placed at risk by being responsible for managing an asset, and that the asset does not create any domestic violence or exploitation issues.

② Appropriateness

Asset distribution programming will only work if the beneficiaries are ready, able, and willing to use the assets, and if this activity will result in a viable livelihood for them. Assessments should include discussions with the potential users of an asset, especially focusing on preferences (such as size or type of equipment) that might prevent an individual from using the asset.

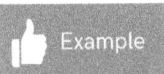

After a conflict, hybrid cows were brought to a community and distributed to the most vulnerable groups, including the elderly, to ensure their food security. A week after the distribution, one grandmother returned her cow to the project office saying that the cow was too big for her to handle and 'mean'. She asked if she could sell it to buy a smaller local cow that she could manage more easily.

③ Impact on vulnerable groups

Ensure the needs of specific groups (women, youth, people with disabilities, the elderly, and disadvantaged ethnicities) are considered and take care not to support cultural discrimination or exacerbate conflict, especially when working with government ministries and departments.

④ Creating a sense of ownership

Livelihood assets are most valuable to people when they feel a sense of ownership over the assets, rather than view the assets as gifts. To increase buy-in, programs may require beneficiaries to provide some level of co-investment in the assets supported by the program, for example by contributing part of the cost of the asset, providing in-kind investments in other materials or labor, or repaying the asset (in cash or in-kind) over time. Care should be taken to avoid individuals becoming over-obligated due to their co-investment responsibility. Where financial institutions (formal or informal) exist, working through them can be one strategy for mitigation, if there is transparency regarding the level of borrowing in the wider community. If target households and enterprises are already investing in assets on their own, then programs should consider complementing these efforts, rather than directly providing assets, as this may actually weaken their coping mechanisms. Communities may be better served by providing broader market interventions that address systemic issues. Regular visits to beneficiaries' homes as they take risks and begin to expand pre-crisis livelihoods or learn new livelihoods can also be an enormous support to their economic well-being. 'Graduation programs' provide considerable evidence of the value of regular contact with households as they try to improve their situation.

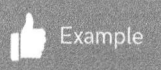
Example

A program in northern Uganda planned to build six warehouses for communities and asked for applications from targeted communities. The applications asked for details on group composition, warehouse governance (once built), and contribution of the group to the costs of the warehouse. It did not require the funding contribution, but said this would make the proposal 'more interesting to judges'. One group that eventually won the competition and had a warehouse built told evaluators that they were very proud that the warehouse was there because 'of our hard work. Most of the community does not know about the financial inputs from the NGO, they only know about the technical assistance. We told them that we did it ourselves.'

Valuing the co-investment

One common technical question is how to determine the level of co-investment to request from asset recipients. This requires an understanding of the economic situation of the potential recipient (either specific or generalized across a community), which should be drawn from surveys or other livelihood data, plus good judgement. A frequent starting place is a 50/50 split of costs, but that split does not consider the level of risk faced by the recipient of the productive asset. In general, the lower the risk faced, the more the recipient should contribute. While the expenses of rebuilding after a crisis should be considered, as stated earlier, there is a relationship between the amount invested and the level of commitment. The beneficiary should contribute at a level that will ensure their commitment to using the asset productively, without undermining their ability to recover from any shock recently faced.

 Risk assessment

Asset programs can inadvertently increase a range of risks, including physical danger from carrying and storing cash and assets, social conflict between beneficiaries and other community members, and long-term problems due to managing enterprises that are not viable. Programs should take these risks and sensitivities into account and reduce them as much as possible.

> ⚠️ **Cautionary tale**
>
> A cash distribution program for asset replacement announces publicly that target beneficiaries can pick up their cash grants at a tent on the community soccer field starting at 10 a.m. on the following Saturday. People wait in line in the open without any security and exit directly onto a main road. It is very easy to see from the road who receives cash and several of the beneficiaries are robbed in the next few days.

> ⚠️ **Cautionary tale**
>
> In Lebanon the distribution of physical assets to refugees was perceived negatively by the host community. Although the e-card modality used to distribute cash assistance is generally a dignified way to provide assistance, watching refugees line up at ATMs to withdraw their cash served to fuel misperceptions and increase tensions with the host community, whose incomes and livelihoods had also been negatively impacted by the crisis in Syria.

Asset Distribution Standard 2

Asset programming stimulates recovery without undermining local markets

All asset distribution programming, whether for protection, replacement, or livelihood expansion, should stimulate recovery and minimize adverse impacts (both short and longer term) on local economies.

Key Actions

- Use assessment findings to identify ways to deliver assistance through local market mechanisms and support the re-establishment of markets when feasible (see also *Assessment and Analysis Standard 2*).
- Develop communication plans and strategies to ensure asset donations are transparent from the community's perspective.
- Develop clear ownership, governance, and management plans for all significant asset transfers. This is particularly important with transfers to groups, communities, or institutions.
- Plan interventions so they can be transitioned to longer term sustainable activities.

Key Indicators

- Assessments have been used to determine potential impacts of intervention on the local economy over the short, medium, and long term.
- The program has a positive or neutral effect on the overall market system and does not create serious market distortions.
- The household and/or community understands and participates in asset management, when appropriate.

Guidance Notes

① Effects on local markets

Assessments need to examine the effect of asset programs on local markets and consider a range of supply-side issues linked to the purchase and distribution of assets (e.g. how local procurement could affect local availability of the good for other people). Local markets should be used for asset purchase if they can meet basic needs and provide the productive assets that households need. However, these markets must be able to respond to the increased demand that comes from a market intervention without suffering from inflationary pressure, which might increase prices and keep other households out, violating the do no harm principle of humanitarian interventions.

Assessing available stocks, current and historical prices, transportation, and logistical challenges related to increasing the supply of goods will help organizations design an intervention that supports current and future market functionality. Regular market price monitoring of key goods, in addition to those being distributed or purchased, will permit agencies to stop or adapt their programs if evidence of inflation appears.

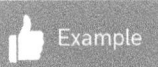

A program buys basic agricultural equipment (e.g. hoes and wheelbarrows) from a local wholesaler and distributes it to farmers to replace their tools lost in floods. This large order gives the wholesaler the cash necessary to restock other merchandise, which increases the availability of farm supplies in the affected area for all farmers.

 Cautionary tale

Following a crisis, the government extensively distributes seed to affected farmer populations. Since everyone now has free seed, the price of seed in local farm stores plummets and the income of local seed producers, importers, wholesalers, transporters, and retailers decreases dramatically. The importers order less seed for the next season, unsure of what the government will do, and the supply of new, drought-resistant varieties is limited because they do not want to hold high stocks if they cannot sell the varieties at an appropriate price.

② Short, medium, and long-term impacts

Many asset recovery and protection activities aim to quickly begin, restart, or expand livelihoods to address immediate household income needs. However, programs should try to assess the long-term viability of different livelihoods, their impact on the broader local market for those goods and services, any effect on local labor markets, and implications for the environment. For programs that involve giving funds or items, organizations must coordinate with other programs (see also *Core Standard 2*). If organizations in the same area take different approaches, particularly if these approaches require more commitment from recipients, there may be conflicts. In certain circumstances, asset distributions can be a link to longer term recovery activities, for example, by enabling micro-entrepreneurs to restart their businesses so they can participate in business development activities. Asset distributions can interfere with recovery efforts if they involve inappropriate distribution chains or targeting. For example, widespread distribution of items purchased externally can weaken attempts to develop local suppliers (potential indirect beneficiaries of the program) of those products.

❸ Transfer modalities

Cash transfers are commonly used in early recovery programming as a flexible and market-sensitive way to address meeting basic needs and protecting assets. Through communication with affected households, programs can help maintain livelihoods and reduce negative coping mechanisms and the sales of productive assets. In addition, programs may involve market solutions to protect or sell assets of declining productivity, for example, by storing crops until prices rise or selling livestock in response to a chronic drought. Other complementary program methodologies may have a direct impact on asset recovery and protection. For example, a financial access program might offer loan guarantees to small enterprises to enable them to access credit and replace assets lost in a crisis and get markets functioning again. When jobs are scarce, cash-for-work programs may enable laborers to earn income while rehabilitating community infrastructure. Many documents provide guidelines and tools for these methodologies – see the Annex for details.

Example

In the Philippines, a randomized control trial for cash distributions after Hurricane Haiyan found that single sum transfers, when compared to three payments (same total amount), resulted in increased ownership of small productive assets. Households receiving the single payment invested more in small livestock (such as hogs, goats, and poultry). Any potential disadvantages of receiving a lump sum did not materialize over the period of the study.

Asset Distribution Standard 3
Productive assets are protected

Preventing the loss of productive assets is done in an appropriate and timely manner, so that households can meet basic needs without resorting to loss, consumption, or sale of assets that would compromise their recovery.

Protecting, slowing, or stopping the loss of productive assets is a first step in helping crisis-affected populations recover. Typically, populations must have stabilized and must be able to meet their most basic needs before organizations can be ready to start replacing lost assets.

Key Actions

- Identify interventions to stop or reduce continued use of negative coping strategies and depletion of existing assets. Learn which assets, if lost, would have irreversible consequences for recovery and focus on the protection of these.
- Determine the approach used for asset protection (for example, cash distribution) based on existing and potential delivery mechanisms, and on the potential impact of that approach on local markets (positive or negative).
- Look for market actors who may be able to support asset protection and work with them to develop innovative solutions. For example, wholesalers who want to protect their supply of inputs may extend credit or advance purchase.
- Monitor needs and coping strategies at the household level. Do this on an ongoing basis and alongside any monitoring of local markets (such as price monitoring); include questions related to supporting items (such as feed for animals).

Key Indicators

- Households are using coping strategies that are healthy, safe, and do not undermine their future potential for earning income. Children are able to stay in school.
- Households retain productive assets during the immediate crisis, or minimize their loss in the immediate aftermath of a shock.
- Market assessments are used to identify market actors that can contribute to asset protection solutions.
- Assets are protected by insurance.

Guidance Notes

❶ Meeting basic needs

Following a crisis, meeting basic needs, such as food, water, healthcare, and shelter, will take priority. Yet, productive assets play a role in providing these basic needs during normal times, and therefore it is vital that interventions begin before households are forced to consume or sell productive assets. Although some of the affected population may have sold assets immediately following the crisis, programs can intervene to prevent further depletion of assets. Special attention must be given to seasonal calendars, as repercussions can be more severe and long-lasting when agricultural interventions are too late. Cash transfers, vouchers, or in-kind support should be targeted at meeting basic needs and designed to encourage protection, maintenance, and, where possible, growth of productive assets.

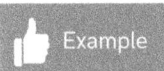
Example

> Immediately following a hurricane, a seamstress sells her thread and fabric so she can buy food for her children. Without these items or money to replace them, her sewing machine is now useless. A program provides cash transfers, so that vulnerable households can purchase food and essential non-food items. The timely transfer of cash prevents her from being forced to sell her sewing machine, allows her to purchase food for her children, and enables her to purchase fabric and thread to restart her business.

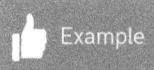
Example

After a flood, a program provides a cash transfer so affected rural households are able to meet basic needs and pay farm laborers. Complementing the cash transfer, the program provides training on proper techniques for digging sustainable irrigation channels. The cash allows farmers to meet basic needs and prevents consumption of seed stock. The training and labor support components decrease farmer vulnerability to future flooding, therefore protecting assets and investments from the impact of future shocks.

❷ Coping strategies

Programs seeking to protect productive assets from the effects of recent crises must acknowledge the realities of the post-crisis environment. While some interventions ease the impact of the shock and decrease the need for negative coping strategies, affected populations may still be forced to alter their behavior to cope with changes. Asset distribution programming should address this reality and discourage coping strategies that endanger productive assets, or are unhealthy or unsafe. Households should keep children in school and avoid removing them to participate in income-generating activities.

Example

After a drought, a comprehensive response program includes a community sensitization component that encourages rural households to maintain seed stocks and emphasizes the longer term importance of planting. Even if households are convinced to keep their seed stock, they may still have to cope in other ways, such as eating less expensive or less preferred varieties of food until the next harvest.

❸ Measure asset retention

Ideally, asset programs can demonstrate that their activities result in higher levels of asset retention and improved resilience to future crises. Many asset programs do not continue long enough to measure this level of impact. Where there are long-term programs helping communities respond to numerous crises over an extended period of time (for example, programs that help communities deal with annual monsoon floods or climate change-related crises), this type of impact measurement is extremely valuable.

 Example

Ongoing program monitoring found that refugees who received a variety of equipment to support microbusinesses were, in fact, selling the items to finance onward travel to third countries. As a result, the program was adjusted to foster joint business ventures between members of the refugee and local host communities. This improved relations between the communities and facilitated registration of the businesses.

Asset Distribution Standard 4
Asset replacement is fair and transparent

When critical livelihood assets have been destroyed, they are replaced in fair and transparent ways, helping households recover and/or strengthen their economic capacity without undermining the local economy.

Key Actions

- Assess the feasibility and appropriateness of continuing or resuming pre-crisis livelihood activities. (Is the livelihood still viable? Is it sustainable?)
- Identify gaps in key assets required to sustainably resume pre-crisis livelihood activities and determine the cost of the required productive assets.
- Match assistance provided to individuals' capacity, skill sets, and market opportunities.
- Recognize that 'equal' is not always fair and seek solutions that will incorporate individual circumstances and vulnerabilities whenever possible.
- Communicate clearly with households, leaders, and communities on the intervention criteria.
- Provide training when introducing new asset types.
- Design short-term interventions with longer term market development and economic recovery goals in mind, whenever possible. Do not undermine economic recovery activities that have a longer term vision.

Key Indicators

- Market assessments are used to determine medium to longer term viability of livelihoods.
- Beneficiaries use assets as intended, with minimal incidence of asset sale or diversion.
- Income, production, or relevant activity-specific indicators return to pre-crisis levels or better and are sustainable at those levels for the medium term (proxy for 'livelihood has recovered').
- Households can articulate the criteria for the program and why they were selected (or not).
- Appropriate training is provided when the intervention involves new equipment or technologies.
- Consideration is made for households that are uniquely vulnerable.

Guidance Notes

1 Respecting household requirements and abilities

Programming interventions must respect individuals' skills, capacities, and ambitions. Asset programming in the immediate aftermath of a crisis or during a prolonged disruption must consider their ability to implement the economic activity targeted by the program in both the short and long term. In addition, programs should consider interventions that are adapted to the changing circumstances of the labor and economic markets. In certain contexts, such as those of large-scale displacement, pre-existing livelihoods may no longer be viable for several reasons. If context assessments indicate this is the case, it may be appropriate to help people diversify into new activities. In such cases, the political context and intercommunity dynamics should be taken into consideration from a do no harm perspective. There is further guidance for these programs in the next standard.

Cautionary tale

After the Indian Ocean tsunami, thousands of boats were distributed to local populations to enable them to restart fisheries. Most boats distributed were small, coastal fishing boats, even though local fisheries previously included many types and sizes of boats targeted at specific fish species and geographic zones. Because of poor targeting and insufficient analysis of usage and impact, many boats were not appropriate and were never used for their intended purpose. Despite this, the large numbers of boats provided still created concerns about overfishing.

Cautionary tale

As a response to the Syrian crisis, certain international organizations based their skills training on the preference of the beneficiaries rather than the real market needs, and trained women in hairdressing and beauty care, though a negligible number of the trainees could make any income from these skills.

2 Transition strategies

Programs in the immediate aftermath of a crisis often aim to support the rapid recovery of households and individuals by simply replacing productive assets, and make limited attempts to improve their long-term situation or add to economic development. Even in these immediate stages, programs should consider the potential long-term effects of the asset program and begin to identify possible linkages to longer term interventions, such as financial services or enterprise development. These programs may be offered by other organizations in the area. (See also the *Financial Services Standards*, *Enterprise and Market Systems Development Standards*, and *Core Standard 2* for more information.) Interventions should focus on supporting, strengthening, and rebuilding the local economy and local actors (see also *Asset Distribution Standard 1*). In post-conflict or natural disaster settings, reviving livelihoods can also be done by investing in rehabilitating some damaged community assets, notably through labor-intensive debris removal or shelter rehabilitation programs. Transition strategies should seek to build on these local strengths and slowly remove external actors as appropriate.

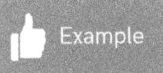
 Example

One organization implementing a cash-for-work program to rebuild infrastructure offered laborers a built-in savings option, where the organization held back a small portion of their salary each week, which would be given to them at a specified end date. This allowed workers to have some funds to meet their family's immediate needs, and a small lump sum to invest once the infrastructure project was completed.

③ Increasing resilience

To fully address asset protection, programs must include disaster risk reduction strategies. At a minimum, programs should seek to reduce vulnerability to future crises, strengthening the impact of the initial post-crisis interventions. Households should be able to protect their assets from the impact of future shocks. Depending on the livelihoods of targeted individuals, these interventions can range from strengthening links to financial services (e.g. insurance or safe savings) to rehabilitating irrigation channels and installing soil conservation structures. Another element of building resilience is to recognize that each household or group is facing a slightly different situation, and blanket asset distributions, although equitable, are not always the right solution. The more that interventions can recognize the unique safety, market access, knowledge, and power issues faced by different groups (e.g. women, disadvantaged ethnicities, very poor households) and then transfer productive assets to strengthen the economic opportunities for these groups, the more resilience will be built.

Asset Distribution Standard 5
Assets expand and diversify livelihoods

Asset distribution programs enable households to take control of their economic recovery, strengthen their future economic potential, and take advantage of new economic opportunities by expanding, adapting, or developing new livelihoods.

Key Actions

- Use market assessments to analyze market conditions to ensure the viability of new livelihoods (see also *Assessment and Analysis Standards*).
- Support people to make informed decisions about the potential of livelihood activities and the need for additional assets.
- Support private-sector and government efforts to introduce new technology and approaches.
- Assess the impact (economic and environmental) of introducing new practices and technologies and share this information with communities and leaders.
- Facilitate access to the complementary services or assistance that households and enterprises will need to utilize these new assets, such as training, financial services, and market linkages.
- Identify key assets that are vulnerable to common, recurrent, or anticipated shocks, and design activities to address or mitigate these vulnerabilities.
- Include measures of medium and long-term viability and sustainability of the economic activities in monitoring frameworks, and share this information with households.
- Facilitate access to regulatory and market information that is relevant to individuals' livelihoods, in partnership with local government if possible.
- Regularly visit targeted households or places of business to determine ongoing needs, link people to additional services, and account for asset wear.

Key Indicators

- Households and/or enterprises are co-investing in new livelihoods at significant levels.
- Households and/or enterprises have increased income due to project interventions.
- Market information is used by project staff and households to make decisions.
- New livelihood activities do not undermine future economic opportunities.
- Households and enterprises have access to appropriate complementary services.
- A plan is in place for the replacement of productive assets that become worn out or are lost due to anticipated shocks. (For example, if the asset is an older milk cow or one that might not produce during droughts, a savings account for replacement might be part of the long-term plan.)
- Where appropriate, sustainability measures are included in program documentation and systems.

Guidance Notes

 **Complexity**

Programs for livelihood expansion are more complex than immediate asset replacement programs (see also *Enterprise and Market Systems Development Standards*). They have the advantage that households are out of the response phase and may more easily participate in program planning and execution. Programs of this type usually require complementary services and technical assistance to achieve their objectives. Because they are subject to continuous changes in the markets, attention to longer term implications is important for successful intervention design and implementation. A systemic approach should be used (see also *Enterprise and Market Systems Development Standard 3*), and programs should assess which supporting services are needed once the intervention is over as well as how the market system will provide that (e.g. access to savings and loans, information on suppliers and buyers, access to appropriate labor). Providing assets on their own is usually insufficient to guarantee impact, and programs should consider a package of services, either before providing the productive asset (such as skills training, followed by provision of a start-up kit) or after it is provided (e.g. coaching and mentoring, business management training).

 Example

A program wants to help rural households sell products in a new market in a town that is quickly expanding. A market analysis shows that the transportation service is too infrequent to enable the households to get their goods to the market. In addition to helping farmers obtain more assets to increase production, the program facilitates conversations between the minibus association and the village groups so the transportation schedule matches the market times.

② Co-investment

Co-investment increases the likelihood that productive assets will be used well. This is particularly important when expanding, adapting, or developing new livelihoods because behavior changes will be required by the household or enterprise if new economic activities are being taken on. Review the guidance under *Asset Distribution Standard 2* when determining the level of co-investment to request from asset recipients. It should factor in risk to households or businesses, and ask them to contribute at a level that will ensure their commitment without undermining their ability to recover from any shock recently faced.

③ New technologies

Introducing technology when providing assets can help individuals adapt to changing environments and opportunities. However, there may be undesired impacts on livelihoods, markets, and the environment if there is not sufficient assessment of the fit with the current situation or follow-through on additional support based on the needs of the individual. The ability of households to use or to maintain new assets is an important consideration. It is vital to assess the income potential of the new assets, in the short and the long term. Training in new technical skills or asset maintenance, as well as investigating linkages to markets for replacement parts and ongoing inputs, may be necessary.

Cautionary tale

Six months after a hurricane, a program gives all microenterprises in the area a US$5,000 voucher to buy equipment or stock for their stores. The microenterprises are free to use the voucher as they wish and are not required to do any planning or invest any of their own funds. Many of the entrepreneurs see this as an opportunity to experiment with new products and types of equipment. However, not all the entrepreneurs know how to manage their new business activities. Within 12 months of the voucher distribution, 50 per cent of the enterprises have exhausted their stocks and/or abandoned the new equipment.

Financial Services Standards

Standard 1
Demand for financial services is understood

Standard 2
Support local supply for financial services

Standard 3
Use existing formal financial service providers for cash transfers

Standard 4
Understand local rules, norms, and support functions

Standard 5
Follow consumer protection regulations

5 Financial Services Standards

Financial inclusion is when individuals and businesses have opportunities to access, and the ability to use, a diverse range of appropriate financial services that are responsibly and sustainably provided by formal financial institutions. Financial services encompass a broad range of products and tools to support and grow assets, along a continuum, which ranges from unconditional and conditional cash transfers to formal financial products. Formal products include loans (credit), savings, insurance, leasing, money transfers (such as remittances and person-to-person payments), and, more recently, mobile money and virtual or e-wallets. At the same time, various informal practices such as savings groups continue to play a critical role in helping vulnerable populations meet their daily financial needs.

Access to a range of financial services is essential for economic recovery efforts, particularly in informal settlements where the population is characterized by poverty, insecurity, and lack of social cohesion. For market-based economies, access to credit is essential for buying food, rebuilding dwellings, paying for medical treatment, and rebuilding businesses after a crisis. In addition, the use of financial services can help individuals, households, and enterprises take advantage of new economic opportunities, create income, and build assets that will enhance individual and community resilience to shocks, allowing them to rebound and recover more quickly. If not provided responsibly, financial services can create new risks in the form of over-indebtedness. As such, responsible finance should be a cornerstone of economic recovery efforts to avoid client exploitation and ensure the recovery of markets that are fair, transparent, and responsive to client needs.

A wide range of service providers – from formal financial institutions (such as commercial banks, insurance companies, non-bank financial institutions, and microfinance institutions) to non-profit organizations and mobile network operators – offer financial services. Furthermore, there are informal financial services, such as community or group-managed savings and credit groups, rotating savings and credit associations (ROSCAs), *hawalas*, and even retail stores that provide goods on credit. After a crisis, it is often the informal providers, predominantly traders and moneylenders, that are the first to start or resume their financial services, as they tend to be more often found in informal settlements and have less stringent lending requirements than microfinance institutions (MFIs) or commercial banks.

An understanding of the actors and market dynamics is critical to supporting financial market systems to serve the diverse and evolving economic and social needs of individuals and households following times of crisis. This comes from understanding both the short-term needs of the recovery and the longer term perspective that allows financial market systems to support poor consumers to rebuild their resilience to future shocks. Since crisis environments are in a continuous state of flux, an ongoing data collection process must be built to continuously inform implementers and enable them to adapt their programming to address this changing environment. Good monitoring helps implementers to uncover the drivers of change in a market, spot leverage points for catalyzing that change, and identify opportunities to create or modify the incentives that are needed to positively influence the behavior of market actors.

For crisis environments, there is no off-the-shelf assessment tool for analyzing the financial sector. However, there are many tools that can be adapted to each situation, provided financial services experts are used in the diagnostic process. (See the Annex for a list of available assessment tools.)

When designing financial service interventions, it is important to ensure they do not replace local institutions or products, or in other ways cause harmful distortions in the financial services market. It is also important to identify the local financial service providers (FSPs) that are able to provide relevant products and services during times of crisis, using the most inclusive and efficient payment systems available. These providers should adhere to responsible finance practices and demonstrate that their products and delivery mechanisms conform to local political, legal, and social norms. Interventions that partner with local FSPs and respect client choice and needs have the potential to create positive effects beyond the intervention. Ensuring there is good market information and adequate training or capacity-building for individuals and providers will lay the groundwork for long-term, vibrant markets.

Financial Services Standard 1
Demand for financial services is understood

The scale and nature of the demand for financial services, including individual needs, preferences, behaviors, and constraints, are understood.

For these Standards, 'demand' includes: 1) those individuals, households, and businesses who are aware of financial service products and are requesting them (recognized demand); and 2) those who are not aware of these products but have needs that the financial service products could meet, and would be likely to use them if they were available and appropriately designed (unrecognized demand).

Key Actions

- Quantify the demand (both recognized and unrecognized) for different types of financial services among target households and enterprises.
- Assess the impact of the crisis on client well-being and/or business and determine existing needs, preferences, behaviors, and use of financial services.
- Identify potential barriers to access and use of formal financial services (e.g. infrastructure; cultural norms/restrictions; regulations, including Know Your Customer rules; and identity documentation requirements).
- Identify complementary needs for formal and informal financial services.

Key Indicators

- Data on the demand for financial services is assessed to determine actual need.
- Distinct client segments are identified and regularly assessed to ensure relevant financial services are provided according to clients' financial capabilities.
- Different types of consumer need (e.g. housing, health, enterprise, household) and times (e.g. lifecycle needs, crisis, rebuilding, investing) are integrated into the demand analysis.
- Pricing and design of products is appropriate to targeted populations.
- Terms and conditions for financial services are easily understood by the target population, and do not exclude them with extensive documentation, collateral, guarantees, or other loan security requirements, such as mandatory savings schemes.

Guidance Notes

❶ Understand scale and nature of demand

Financial services are used by consumers to manage their daily lives as well as to invest in economic opportunities. While formal financial services may be disrupted in crisis environments, the financial needs of individuals are not. This requires individuals to shift from formal to informal services or resort to negative coping mechanisms, such as reducing their food intake or selling off productive assets, if financing options are not available. Crisis-affected households and enterprises require a range of financial services, including savings, credit, payment mechanisms, and insurance. These needs evolve over time. For example, immediately after a crisis, consumers are likely to need immediate access to their savings or may need access to remittance services to receive help from family and friends who are in unaffected areas. Access to inward remittances is a crucial need in a post-crisis or protracted conflict environment. In general, organizations should prioritize inward remittances and cash transfers immediately following a crisis. Next, they should work with FSPs to increase access to savings and insurance. During a reconstruction or recovery phase, FSPs can reintroduce (or expand) loan products based on client demand for access to finance. Emergency loans can help businesses respond to immediate needs, while credit will be needed once a crisis subsides to rebuild businesses and housing and support enterprise development.

Macro- and microeconomic evidence suggests the positive role of remittances in preparing households against natural disasters and in coping with the loss afterward. Analysis of cross-country macroeconomic data shows that remittances increase in the aftermath of natural disasters in countries that have a larger number of migrants abroad. Analysis of household survey data in Bangladesh showed that per capita consumption after the 1998 flood was higher in remittance-receiving households than in others. Ethiopian households that receive international remittances seem to rely more on cash reserves and less on selling household assets or livestock to cope with drought. In Burkina Faso and Ghana, international remittance-receiving households, especially those receiving remittances from high-income developed countries, tend to have houses built of concrete, rather than mud, and greater access to communication equipment, suggesting that they are better prepared against natural disasters.

(Excerpt from *Remittances and natural disasters: ex-post response and contribution to ex-ante preparedness,* Global Facility for Disaster Reduction and Recovery (GFDRR))

❷ Assess demand regularly

Crisis environments are very dynamic, particularly over prolonged periods. This requires FSPs to understand the different segments of the market (e.g. youth, women, farmers, extreme poor) and to stay responsive to individual needs, providing products and services that meet rapidly evolving market and environmental conditions. Financial service providers should regularly seek feedback from individuals and scan the market for changes. Doing so requires them to integrate feedback mechanisms into their operations. It may also require FSPs to undertake ongoing market assessments.

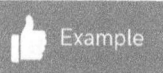

MFIs operating in tsunami-affected areas of Sri Lanka reported larger total savings balances in March 2005 than in March 2004. An assessment in Batticaloa district showed that around 35 per cent of cash grants and cash-for-work payments that were received were deposited in MFIs.

❸ Adjust product design and delivery

Crisis environments may require creative solutions to product design and delivery mechanisms to adapt to the general instability and ongoing economic disruption. After a disaster, individuals are often unable to repay loans according to a pre-disaster schedule. The most common response from financial institutions is to restructure or reschedule loans. Restructuring should take place immediately after a disaster and should be limited to geographic areas heavily affected. Financial institutions should meet with individuals to evaluate their post-disaster repayment capacity. Decisions to reschedule or restructure loan repayments should be based on a detailed assessment of the individual's temporary loss of income. Refinancing, which usually involves replacing a previous loan with a larger one, can be the best option for individuals who have lost their productive assets in the disaster and need a larger loan to replace these assets. Again, a detailed assessment of individual losses is crucial to determine whether to refinance and, if so, the amount and conditions of the loan. Writing off loans should always be a last resort as it can reduce the financial institution's capital and worsen credit repayment culture. After a disaster, financial institutions may consider changing their normal policy to allow more time before loans are written off, in accordance with local regulations. Financial institutions must be prepared to write off loans that cannot be collected due to the death, permanent disability, or disappearance of individuals.

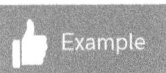
Example

In Syria the Aga Khan Agency for Microfinance offered a secure place for individuals to store their savings and branch-to-branch transfers and withdrawal services during periods of internal displacement. Individuals were also offered the option to use their savings as collateral for a loan, at a proportion that varied with their circumstances. They could also withdraw from their fixed-term deposits without penalty, at any time. First Micro-Finance Syria (FMFI-S) extended business reconstruction loans to current clients who had experienced partial or significant loss of business assets during the crisis but demonstrated the capacity to continue to generate revenue. In addition, the FSP determined whether a client was eligible for loan rescheduling, including an assessment of asset damage and cashflow projections. Rescheduling was done on an immediate and case-by-case basis in order to avoid portfolio contamination. A standard six-month grace period allowed individuals to use loan funds productively to rebuild capacity. FMFI-S allowed loan forgiveness in cases where the client was unable to repay the loan due to death or irrecoverable loss of livelihood. Loan forgiveness was applied on a case-by-case basis and was confidential to avoid moral hazard and other concerns.

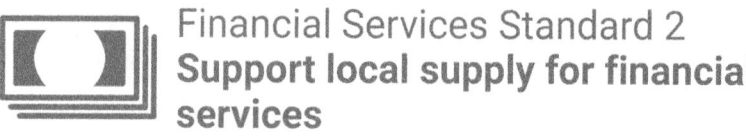
Financial Services Standard 2
Support local supply for financial services

Wherever possible, partnerships are made with local market actors that have in place the capacity, scale, and resilience to continue to deliver financial services in times of crisis.

Key Actions

- Identify the range of trusted local market actors (formal and informal), their geographic coverage, and degree of functionality.
- Assess internal systems of market actors for their risk management, cash handling, reporting, contingency planning, and liquidity management.
- Avoid recreating delivery mechanisms when the capacity for the desired intervention already exists in a pre-crisis institution or market actor.
- Assess and adapt the distribution networks and delivery mechanisms for financial services to improve accessibility, security, and efficiency of delivery.
- Promote linkages, wherever possible, between relief efforts and long-term access to financial services to support sustainable livelihoods.

Key Indicators

- Wide geographic distribution networks and appropriate delivery channels allow financial service providers to effectively service individuals. (Delivery channels may include mobile money, agents, or debit cards.)
- Financial service providers have established relationships in the affected communities.
- Disaster management and contingency plans exist to ensure institutional resilience in times of crisis.
- Financial and operational risk management policies and procedures are in place to minimize institutional risk in times of crisis, including loan rescheduling policies.
- Policies and procedures are in place to separate mainstream financial services from relief and rehabilitation activities, carried out in partnership with humanitarian organizations.

Guidance Notes

 Potential partners

Financial service providers (e.g. non-bank financial institutions, specialized microfinance banks, MFIs, mobile network operators, self-help groups) serving poor and low-income populations have an important role to play in responding to disaster, restarting the local economy, and supporting livelihoods. Because of the profile of their individuals or members, they are already operating in more remote areas of the country (which are more vulnerable to disaster and shocks) through branch or retail distribution networks. They have established relationships and a base of trust with households and the community that can be leveraged during times of crisis. FSPs also have a good understanding of client needs, priorities, and limitations. Wherever possible, efforts should be made to assess the capacity and reputation of potential FSP partners on a periodic basis (at least once year) to ensure prepositioning agreements are in place to minimize response time during times of crisis. The Mix Market is a good source of information on MFIs working in affected areas. Any analysis of the suppliers of financial services should look at their offerings before, during, and after the crisis. It should unpack the provider's ability to respond to and withstand future crises. Furthermore, it should aim to uncover the underlying incentives that drive the behavior of FSPs and the dynamics between them. An understanding of informal providers requires drawing information from demand-side research to help identify what consumers are using and additional analysis to unpack product, pricing, and other data directly from the informal supplier.

❷ Institutional preparedness

FSPs in areas affected by frequent crises need to be prepared for a recurrence of conflict or disaster. Policies and procedures for crisis preparedness and response should be documented, with annual revisions reviewed by staff and the board. Training programs should be in place for staff operating in crisis situations, with systems to ensure the safety and retention of staff. Similarly, the organization's management information system should be designed to withstand disaster, with a clear onsite and offsite backup process, along with other documented procedures on system operation in the event of a crisis. This policy should ensure protection of client information and the institution's access to data. Additionally, at a time of crisis, an institution's ability to manage its liquidity (meet all its payment obligations on a timely basis) may be compromised. Individuals may save less, stop saving all together, withdraw savings, miss loan payments, or request supplemental emergency loans. All these strongly influence the amount of cash moving in and out of an FSP and, for unprepared organizations, may result in liquidity shortfalls. Providers of financial services in crisis-affected environments need to be aware of and anticipate this volatility so they are able to offer options, including mobile money, that help individuals through these situations without compromising the institution's long-term viability.

❸ Delivery channels

Providers should evaluate and select the most appropriate and convenient delivery channels for their services. When feasible, they should leverage technology to save time and costs, while increasing privacy and security. Development partners can assist by reviewing the capacity of existing infrastructure, including agent networks, mobile money, merchant payments, and debit cards. It should be noted that poor and vulnerable populations often face difficulties in using new digital payment systems due to lack of digital literacy, income, and residential status. Therefore, every effort should be made to ensure payment services are inexpensive, accessible, and transparent for individuals in order to ensure comprehension and usage of products.

❹ Staff and client security

Ensuring the security of staff and clients is important in any recovery activity, but it is particularly crucial in financial service interventions, due to the large volume of cash that is handled and the level of confidential client information. Organizations should explore options for the use of mobile money or other forms of electronic transfers, either through mobile network operators or formal banking services, when available. In some cases, these may not be easily available or functioning immediately after a crisis. Organizations should expect recurrent periods of high risk and have policies in place to minimize the potential danger to individuals and staff in terms of managing cash and protecting client information. (See also *Sphere Handbook Core Standard 6: Aid Worker Performance* for more information.)

❺ Ongoing assessment of client needs

Clients of financial service institutions may be affected by crisis in different ways. Some may temporarily or permanently have their ability to repay loans affected and therefore need loan rescheduling or access to their savings. Providers of financial services in crisis-affected environments need to understand and anticipate this volatility, and offer options that help individuals through these situations, without compromising the institution's long-term viability. Organizations should make ongoing assessments of client needs through specialized recovery agents trained to handle non-performing loans. Disaster loan funds, standardized policies for loan restructuring, and credit life insurance to protect against enterprise failure can also help protect individuals' remaining assets and increase their resilience in the face of the disaster.

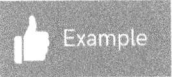
Example

Organizations that want to help individuals negotiate crises might decide to make clients' savings available as soon as possible, without penalty; provide additional capital; reschedule loan payments; drop penalties for late payments for a fixed period; or renegotiate loans to make them interest free. Although writing off a specific loan is an option for the financial institution, maintaining the expectation of repayment is crucial to the organization's ability to continue lending money and thus clients' access to future loan cycles.

6 Access to long-term services

The delivery of financial services is complex and requires a commitment to providing these services over the long term. Service providers should have adequate technical, institutional, and financial capacity. They should be committed to delivering services according to good financial service practices. Successful financial service provision requires financial oversight, accounting expertise, auditing, good governance, strategic planning, and other commitments. Reliability of access over the long term is a key characteristic of good financial services. It is important that any local FSP with which a partnership arrangement is made has a permanent presence in the community and has demonstrated long-term commitment (e.g. investment of funds, time, and specialized expertise) to continuing to provide services after initial response and recovery efforts. If this commitment does not exist, it is best to work with alternative partners, such as community-based financial services that are 'owned' and managed by community members themselves, based on their own savings mobilization (such as ROSCAs and savings groups) or volunteer networks built up over time, or focus efforts on non-financial services.

 # Financial Services Standard 3
Use existing formal financial service providers for cash transfers

Humanitarian actors deliver cash transfers through existing formal payment mechanisms and systems of social protection to increase efficiency and safety.

While financial service providers can play many different roles in an emergency response (as discussed in *Financial Services Standard 2*), this standard refers specifically to using formal FSPs for cash transfers.

Key Actions

- Identify the range of trusted local market actors, their geographic coverage, financial capacity, inclusiveness of payment systems, and degree of functionality.
- Determine if existing safety net or social protection schemes can be used to scale up the delivery of emergency transfers.
- Assess the financial practices and payment behaviors of recipients, including existing access to and usage of payment systems.
- Avoid recreating delivery mechanisms when the capacity for the desired intervention already exists in a pre-crisis institution or market actor. Partner with existing payment providers wherever possible to minimize costs and increase safety of service delivery.
- Promote linkages wherever possible between relief efforts and long-term access to financial services to support sustainable livelihood.

Key Indicators

- Capable financial institutions offer geographic coverage in the zone of intervention.
- Payment systems are robust and resilient enough to reach targeted populations quickly after a crisis.
- Payment systems are efficient and transparent (i.e. merchants can accept digital payments, a network of agents can manage their liquidity, and network connectivity is sufficiently reliable for point-of-sale (POS) devices and mobile phones).
- Emergency response targeting includes the caseload for social protection, and sufficient intra-agency coordination is in place to allow for social protection transfer services to be used for humanitarian response.
- Cash transfers are linked, wherever possible, to longer term access to financial services (e.g. mobile wallets, universal bank accounts, and savings products that are cost-effective and provide value to individuals).

Guidance Notes

 Introduction to cash transfers

While in-kind goods (such as food, clothes, and blankets) still comprise most humanitarian assistance, an increasing number of organizations have begun delivering assistance through cash transfers and vouchers. Delivering cash transfers through digital channels offers many benefits for humanitarian actors, including minimizing the costs (including leakage) and safety concerns related to cash transport. Digital delivery can benefit transfer recipients, providing convenience and choice (the transfer is immediate and recipients can choose when and where to spend it); safety (the recipient does not have to hide the cash anywhere); privacy; and dignity (more difficult for other members of the community to know who is receiving assistance and who is not). While digital transfers can provide significant benefits compared to physical cash, program assessments should nevertheless consider individuals' and households' familiarity with and capacity to operate digital delivery channels (e.g. mobile operating systems and cards with personal identification numbers (PINs)). Additionally, digital delivery requires infrastructure that is not necessary for physical cash disbursement, such as merchants who can accept digital payments; a network of agents who can manage their liquidity; and reliable network connectivity that can access POS devices and mobile phones. A wide variety of tools on cash programming are available from The Cash Learning Partnership.

② 'Piggyback' on existing social protection plans

Social protection schemes provide support to individuals who experience chronic poverty and vulnerability. This support is regular and predictable, which helps to protect against shocks, and promotes the buildup of assets. These individuals become even more vulnerable when a crisis hits, and social protection schemes are increasingly being viewed as a mechanism that can be leveraged to increase cash transfer levels during times of crisis. While examples of partnerships between humanitarian organizations and governments, certain criteria need to be in place for successful scale-up: 1) the caseload for social safety-net schemes and emergency response must be the same because practice has shown that it is easier to scale vertically (e.g. increasing the level of benefit) than horizontally (e.g. increasing the number of beneficiaries); 2) there needs to be a high level of interagency coordination between humanitarian and governmental development agencies and this requires close links with a range of government departments; and 3) the selection of targeted groups or individuals needs to be objective and needs-based, not influenced by the political motivations that underpin many social safety-net programs. Design features such as pre-registration and electronic transfers can further assist emergency response.

③ Linking cash transfers to financial inclusion

Organizations can use digital cash transfers to link recipients with formal financial services, including savings, payments, remittances, loans, and insurance. How this linkage takes place (and for which services) will depend on program objectives and local context. In countries with highly developed mobile money platforms and broad network access, cash transfer programs should leverage the existing system and work with mobile network operators to provide direct transfers into recipients' mobile wallets. Cash transfer interventions should take account of the Know Your Customer rules and consider whether it is possible to open accounts for all cash transfer recipients, including women, who often lack the necessary forms of identification. An account at a formal institution will allow recipients to receive remittances from abroad more easily and at lower costs, often representing a critical lifeline for households in post-crisis environments. Connecting transfer recipients with financial services will require an assessment of their financial experience, behaviors, and preferences, as well as education and training, to ensure that the payment system is not only valuable, convenient, and secure, but will also provide them with a pathway to greater resilience.

Example

Kenya's Hunger Safety Net Programme is an unconditional cash transfer program that aims to reduce poverty, food insecurity, and promote asset retention and accumulation in poor households in northern Kenya. In its pilot phase it used a private-sector payment provider (Equity Bank) and a biometric smart card to make regular, electronic cash transfers to 496,800 individuals. It was implemented under the Ministry of Northern Kenya, with NGO and private-sector implementing partners. The program was later scaled up to add a further 100,000 individuals and link an additional 272,000 individuals with a fully transactional bank account and bank card that could act as a scalable safety net in times of crisis. The project found that these alternative technologies for making social safety net payments were successful in overcoming specific economic and geographical implementation challenges, such as remoteness, thin markets, and high levels of mobility, and could be scaled up in times of crisis to deliver recovery payments to sub-counties badly affected by drought. Less than two weeks from a trigger event, nearly $2 million was transferred electronically into the bank accounts of an additional caseload of more than 90,000 affected people.

Cautionary tale

Past experiences have demonstrated that making the link between cash transfers and financial services has not always been successful. Recipients often cash out their transfer entirely and never use the product again (whether a prepaid card, account, or mobile wallet), possibly because they do not find the service valuable or sometimes because they are concerned they will no longer be eligible for support if humanitarian organizations see them storing money. In some cases, though, uptake has been much higher and recipients have chosen to store a portion of their cash transfer in their e-wallet or account. Program and product design must consider the client's needs and behaviors to ensure the offering is relevant and valuable for the user.

Financial Services Standard 4
Understand local rules, norms, and support functions

Interventions show an understanding of political, legal, and social norms, and strengthen supporting functions for financial services.

Key Actions

- Understand the interrelationship between the political economy and your intervention (see also *Core Standard 4: Guidance notes*).
- Identify and understand relevant financial regulations.
- Ensure interventions conform to existing regulations or obtain the necessary exemptions to operate.

Key Indicators

- Financial products adhere to local laws, regulations, and customs.
- Data protection mechanisms – in place for collecting, processing, and sharing beneficiary information – conform to current regulations, laws, and international conventions.
- Supporting functions exist to reduce transaction costs, improve market information, and enable access to capacity development for both providers and consumers.

Guidance Notes

 Local laws, regulations, and customs

One of the defining characteristics of many crisis environments is the absence of rule of law. Many implementers use this vacuum to design and implement interventions which may have worked in other contexts, but are not supported by existing laws and regulations. For example, interventions that involve savings or insurance schemes in environments where there are restrictions on financial intermediaries or the provision of insurance, will have no chance of long-term sustainability. It is important that interventions are designed using an informed analysis of existing local laws, regulations, and customs. Should the intervention require a change in regulations or changes in customs, then an explicit strategy linked to how these changes will be facilitated needs to accompany the intervention. While seeking exceptions to existing regulations may be feasible in some contexts, organizations must ensure these exclusions do not lead to monopolistic behaviors or an un-level playing field for providers.

 Example

In northeastern Nigeria, traders were not comfortable paying out electronic cash grants for emergency food because they felt it was not in line with the Koran. Instead of a cash grant payout, the traders agreed to use electronic food vouchers in their areas as a compromise.

② Supporting functions

Given the nature of financial services, information plays an important part in enabling the exchange between financial service providers and consumers. FSPs contribute important market information – through credit bureaus, consumer research, public campaigns, and disclosure regulations – that helps consumers to understand financial products and make informed decisions on providers. Other supporting functions include training and capacity building for consumers and providers, which can reduce the knowledge and behavior barriers on both sides. Interventions that aim to facilitate financial services for crisis-affected populations should understand the type of supporting function that is needed to improve a specific financial sub-market (e.g. insurance, payments, savings). At the same time, interventions that directly address supporting functions should be designed using a systems approach (see also *Enterprise and Market System Development Standards*).

③ Working with refugees/IDPs

Refugees and IDPs are population segments that are particularly affected during crises. In some contexts, the financial services that are available within the host communities may work in a suitable way for refugees and IDP populations; understanding the demand side and broader regulatory and social issues that affect these segments is critical for supporting their needs. In many countries, refugees do not have the right to work, which will impact their demand for and ability to use some types of financial service. In some countries, refugees are not allowed to open a bank account. In countries where these rights may exist, refugees may not have the documentation that is required to obtain a job or open a bank account. The vulnerability of refugees can vary quite dramatically, depending on when they left, and how they left, their homeland. Some refugees may have access to assets they could take with them, whereas others may have nothing more than what is on their backs. Since refugees often put strain on the resources of host communities, understanding the social context in which the refugees reside is important in designing interventions that do not fuel further tensions with host communities.

While IDPs are often seen in the same light as refugees owing to their status as displaced people, the regulatory and social constraints they face are often quite different and need to be understood by implementers and FSPs interested in serving this segment. Unlike refugees, IDPs are in their homeland, but they have been forced to move due to conflict or natural disaster. While IDPs may have the same rights as other residents in their host communities, they have often suffered traumas that require specific attention. And while they may have the necessary identity documentation to access employment opportunities or financial services, they often lack collateral or social networks and can be susceptible to discrimination by host communities in the same way as refugees. It is incumbent upon implementers and FSPs to understand the specific needs of these population segments, and the broader contexts that will define what refugees and/or IDPs can do regarding employment, self-employment, and access to financial services.

Financial Services Standard 5
Follow consumer protection regulations

Interventions adhere to universal standards for social performance management, treat individuals in a responsible manner, and respect all local consumer protection laws and regulations.

Key Actions

- Ensure financial products provided do not create over-indebtedness among the target population.
- Ensure partners have a demonstrated commitment to good practices for pricing transparency, fair and respectful treatment of individuals, and mechanisms for complaints resolution.
- Ensure delivery channels are well understood by individuals and provide for the safety and privacy of data.

Key Indicators

- Financial service providers have sound policies and well-documented processes in place to ensure credit is extended only to borrowers who can repay their loans and who are not already over-indebted.
- The interest rates, prices, fees, and terms of all financial products are explained to individuals in a manner that is transparent and understandable, both orally and in writing.
- Debt collection practices are respectful and not coercive or abusive.
- A code of ethics is in place and applied to combat corruption or abuse of individuals.
- Client complaints are responded to and resolved in a timely manner.
- Training on new technology or delivery mechanisms is provided to recipients to ensure they are able to effectively access and utilize services being offered.
- Client data is kept secure and confidential; financial service providers respect individuals' privacy, adhere to local laws and regulations, and do not use or distribute individuals' data without their permission.

Guidance Notes

① Avoiding over-indebtedness

While practice has shown that individuals can often use loans immediately following a crisis, it is not an appropriate intervention for individuals who are not economically active. Well-timed and properly structured loans will be critical in assisting economically active individuals to rebuild their lives. This requires aligning all steps in the credit process – from design to appraisal, monitoring, and reporting – to ensure credit is needed and that the resulting debt does not exceed an individual's capacity to repay. Sharing client credit histories with other financial service providers, where appropriate and legal to do so, can help avoid making multiple loans to the same client, further reducing the chance of over-indebtedness.

② Adherence to good practices

Individuals need to understand the financial services on offer, including the terms, conditions, and processes, as well as their rights and obligations under these contracts. Providers should treat individuals fairly and respectfully and not discriminate, have in place effective mechanisms for complaints and problem resolution to address individual concerns as they arise, and work to improve products and service offerings.

③ Providing training on new technologies

Digital financial services provided through mobile phones, wire transfers, cash machines, and so on, hold great promise for the humanitarian community: creating efficiencies, reducing costs and potential leakages, and increasing accuracy of targeted interventions, particularly when it comes to cash transfers. However, emerging technologies involve new, and often complex, interfaces and complicated processes that can intimidate users and reduce uptake and effectiveness. This is particularly the case for poor and vulnerable populations, who have low levels of literacy and are often new to, or uncomfortable with, digital payment systems. As such, it is important that recipients of digital financial services receive thorough training in, and communication about, new digital delivery mechanisms before and after roll-out. As a minimum, individuals should know their payment amount and frequency; how the system and payment mechanism should work; and where to go if they face problems.

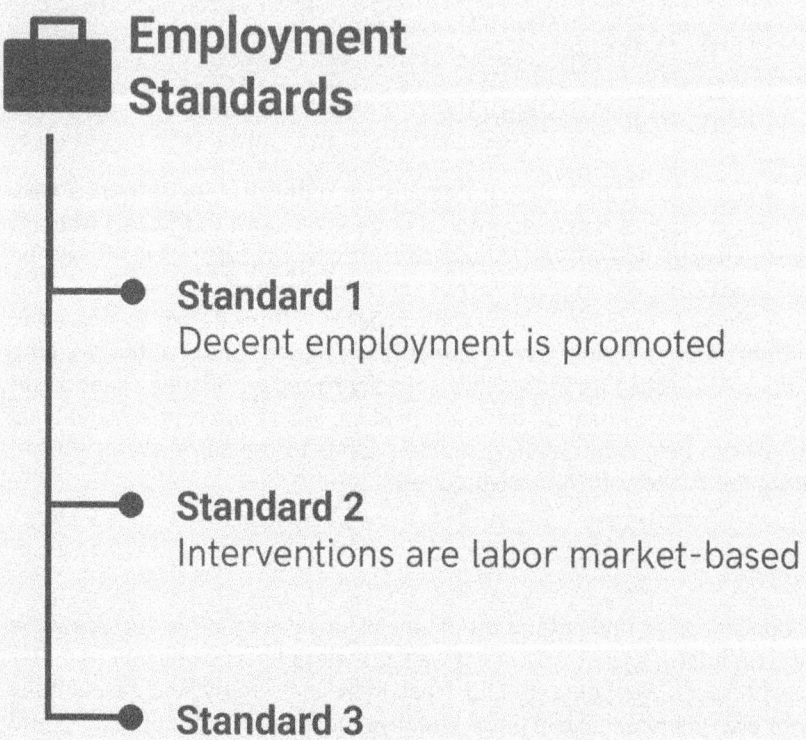

6 Employment Standards

The Employment Standards relate to activities that prepare individuals for work or create jobs through humanitarian and economic recovery projects. Interventions should focus on promoting decent and safe working conditions that allow individuals to earn a living wage. Employment programming should also consider labor market realities and sociocultural context. Additionally, such programs should allow for routine evaluation and integration of revised programming that is responsive to ongoing labor market assessments. Employment programming should encourage ongoing skills building and consistent employment. The importance of productive employment is highlighted by the United Nations' Sustainable Development Goal (SDG) 8, which aims to 'promote sustained economic growth ... [and] to achieve full and productive employment, and decent work, for all women and men by 2030'.

Employment activities may include:

- business skills building or vocational training
- setup of on-the-job training opportunities and establishment of apprenticeship programs
- career counseling and job placement
- temporary employment to rebuild infrastructure, such as cash-for-work programs
- creation of trade membership associations and workers' cooperatives
- partnership with private-sector employers in design of interventions and placement of graduates
- creation of jobs through investments in enterprises
- training of employers on safe and decent workplace standards.

Employment Standard 1
Decent employment is promoted

Employment opportunities deliver a fair income, provide physical, emotional, and social protection within the workplace, and allow prospects for personal and professional development.

For more information on decent work, see the Glossary and International Labour Organization (ILO).

Key Actions

- Assess national and local government policy as well as the informal norms and customs that govern the labor market.
- Incorporate the ILO's Rights at Work standards for vulnerable populations into project planning where gaps exist in local policy.
- Advocate for fair wage levels based on current market realities for targeted employment sectors. (Fair wage rates must resolve any differences between the prevailing wage rate and a living wage that allows employees to meet basic needs.)
- Work with enterprises to ensure management understands decent and equitable work standards and promotes inclusive hiring.
- Provide technical support to enterprises committed to upgrading the physical, emotional, and/or social protection of their workers.

Key Indicators

- Work promoted meets quality, safety, and do no harm international standards (such as ILO, Sphere, SDG/World Bank, UN Climate Change Conventions, child labor laws).
- Programming supporting refugees and internally displaced people Internally Displaced Persons (IDPs) is aligned with UN and ILO fundamental right-to-work standards regarding displaced people (UN, 1948: Article 23; ILO, 1998).
- Interventions are inclusive and program targeting takes into account the unique needs of vulnerable populations to ensure equitable and fair access to decent employment.
- Job creation activities and private partnerships uphold and promote quality of employment.
- As necessary, improvements to programming and work conditions are recommended, supported, and regularly assessed.
- Work provides a living wage that, at a minimum, allows people to meet their basic needs. The wage rate also takes into account local labor rates and the specific purpose or goal of programming.
- Employers understand the standards of decent and equitable work and use inclusive hiring practices.
- Where humanitarian organizations are the employer, they coordinate with other agencies and the private sector to ensure that labor rates are consistent and do not distort the market (see also *Core Standard 2*).

Guidance Notes

1 Quality of employment

The following factors should be considered in determining whether employment opportunities are decent:

- The *level of remuneration* is appropriate and allows for an adequate standard of living, taking into account local labor rates and any coordination of wage rates by international agencies. Payments for waged labor should be prompt and regular. It is important to remember that wage rates above market rates for similar work will draw workers away from local private-sector and agricultural actors, and lead to longer term unemployment, business closures, and other negative effects.
- Procedures are in place to provide and promote a *safe, secure working environment*. This includes considerations of gender, child labor laws, ethnic background, people with disabilities, the elderly, and other vulnerable populations (see also *INEE Minimum Standards, Education Policy Standard 1: Law and Policy Formulation*).
- Employment is *respectful of human dignity* and offers opportunities for personal and professional development. For instance, program strategies should consider protection, respect minimum work ages, and not undermine people's responsibilities to care for their household (see also *Sphere Handbook, Food Security – Livelihoods Standard 2: Income and Employment*).
- Employment and job training opportunities are *equally accessible* to women and men, to those of different religious, ethnic, and political backgrounds, to youth who are old enough to work, to the disabled and the elderly, and to all groups in a community, including (where relevant) host community members, IDPs, refugees, returnees, and demobilized combatants. This does not necessarily mean that all targeted groups will be trained in the same vocations. It simply means that they have equal opportunity to access employment and training interventions.
- Employment *respects local laws and customs*, where these do not contravene international standards and the quality criteria listed here.
- Program design *promotes formal employment* when possible.

As part of its employment intervention, an NGO ensures apprenticeship placements with a number of local employers for graduates of its job training program, to help them gain on-the-job training that may turn into full-time employment. Each apprentice is issued an ID badge identifying them as an 'official' apprentice, and on the back is a confidential hotline number for them to call if they encounter any discrimination or harassment issues on the job.

② Operating environment

Project design considers local and national government policy, as well as the formal and informal norms and power structures which govern the labor market. This context should be considered when conducting the labor market assessment. When policy or social norms violate international standards for employment, advocacy to educate market actors and address these issues should be considered in the project design.

Long-term refugees are constrained by government policy to remain in a camp and are forbidden from working outside the camp setting. Nonetheless, a vocational training provider in the camp begins offering car mechanic training, though there are few cars in the camp. When refugee youth graduate from the training, they cannot find any employers with a need for their mechanic services and remain unemployed.

③ Human capital development

In identifying viable employment options, the skills, support, and training necessary to meet employment requirements – and the feasibility of offering these – should be considered. Programs should also assess the need to provide training in life skills, such as literacy, leadership, and communication, in addition to providing the appropriate psycho-social support.

④ Targeting

Employment needs are likely to be much greater for particularly disadvantaged and vulnerable people (such as people with disabilities or women in contexts where they have limited freedom of movement), yet it can be challenging to include them in standardized employment schemes. Alternative access to employment will always need to be facilitated for these groups.

 Example

> An employment training program is set up by an NGO to be delivered through a community center in an urban area. The program seeks to improve employability for both men and women, yet attendance by women is low. After conducting focus group discussions, the NGO realizes the women must watch their children when classes are scheduled, so it arranges to provide child care during classes so that more women may attend.

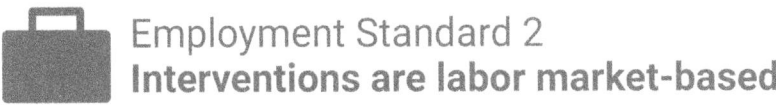

Employment Standard 2
Interventions are labor market-based

Employment interventions are based on current market conditions, future opportunities, and the employment competencies present in a population.

Key Actions

- Conduct pre-crisis labor market assessments to increase preparedness and contribute the data to program design during humanitarian response and economic recovery (see also *Assessment and Analysis Standards*).
- Conduct ongoing analyses of the supply and demand of labor, products, services, and safety and security; adjust programming to account for changing market and labor conditions (see also *Enterprise and Market Systems Development Standard 3*); and program adequately to meet the needs of mobile populations.
- Complete comprehensive market-mapping assessments to inform humanitarian response and program design, and identify labor market shocks. Update mapping regularly throughout program implementation and in the post-crisis context so that findings may be incorporated into programming.
- Identify norms (policy or sociocultural) that create limitations for the targeted groups. Determine whether both formal and informal employment options should be considered for program design.
- Work with private-sector actors to identify jobs in demand (needed in the economy) and design programs according to identified needs. Look for consistent, sustainable job opportunities that interventions can be built around.
- Ensure the employment sectors chosen by the project do not worsen contextual, conflict, or environmental issues.

Key Indicators

- Pre-crisis and regular labor market mapping analyses provide data to inform employment programming (see also *Assessment and Analysis Standards*).
- Interventions are evaluated and adjusted regularly, based on new data and changing labor demands. Evaluations include, but are not limited to: employment training, wage rates, and private-sector partnerships.
- Program beneficiaries are usually able to find, obtain, and hold on to employment that provides a wage which allows them to meet basic household needs.
- Enterprises recognize future opportunities in the market and invest in building the skills and competencies of their employees.

Guidance Notes

① Market assessments

Program strategies and interventions should be informed by research on labor market trends. Research to determine the demand for labor involves examining: 1) current and emerging sources of employment in the local economy, 2) the potential for trainees to be hired on completion of employment programs, 3) competence levels and certifications required by trainees to enter the labor market, 4) any gaps in labor market skills and knowledge, 5) existing systems and resources for workforce development in the public and private sectors, 6) how gender, caste, age, or other biases affect hiring and employment practices, 7) prevailing wage rates for different sectors and sub-sectors, 8) seasonal effects on labor demand (both rural and urban), 9) migration patterns that affect labor supply, and 9) how unanticipated events (such as climate events and political unrest) might be mitigated by DRR planning in labor markets. (See also *INEE Minimum Standards, Teaching and Learning Standard 1.*)

Pre-crisis market mapping, when possible, should be conducted and used to build resilience into a community, including identifying existing employment experience and competencies that may be leveraged post-crisis (see also *Assessment and Analysis Standards* for sample assessment tools).

② Labor market and employment monitoring

Labor market assessments and employment monitoring should take place frequently to ensure interventions remain relevant in a rapidly changing post-crisis context. Assessments will help determine if local labor markets have been influenced by changes in the availability of goods or services, political stability, or climate events, and can ensure the ongoing appropriateness of job training, employment opportunities, labor wages, and the safety and security of beneficiaries. Post-program monitoring of learners from vocational training programs can provide valuable feedback for program design. (See also *INEE Minimum Standards, Analysis Standard 3.*)

Private-sector involvement

The private sector must be informed and involved in planning to ensure that the skill sets developed meet market demand, and to provide opportunities for placement, mentoring, and ongoing workforce development when the intervention is completed. When policy or practical limitations on target groups are identified, informal employment options may be considered, but private-sector actors should still be involved if possible. Private-sector partners providing training, apprenticeship, mentoring, employment, and other interventions should be routinely evaluated to ensure their programming and environments meet standards for decent employment (see also *Employment Standard 1*) and, ideally, a confidential feedback loop should be made available for trainees.

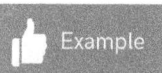

> During the inception phase of a project, an NGO partners with a large company to develop the business and technical skills curriculum, with the company contributing a number of insights and, in particular, helping to design modules aimed at improving an individual's business communications and customer service skills, skills which it had identified as lacking in the employee pool.

Employment Standard 3
Job sustainability is supported

Employment interventions support job sustainability and longer term employment opportunities. In dynamic contexts, these sustainability goals may be achieved by a stepped approach, working through short and medium-term interventions.

Key Actions

- Cultivate partnerships with the private sector to ensure employment opportunities are aligned with market needs; this makes the private sector more likely to create longer term opportunities for targeted groups after the intervention has ended.
- Focus short-term employment interventions on skill sets in demand, or on sectors with strong growth potential, to help individuals benefit from their experience once the activities end.
- Design employment interventions that consider the capacity and preferences of targeted individuals or groups.
- Provide additional support to groups that are not fully integrated into the labor market (such as women, people with disabilities, and youth) and look for creative solutions to cyclical employment issues (such as migration).
- Help ensure that training certification and employee registration mechanisms exist through a local state or non-state entity.

Key Indicators

- Public and private entities are active partners in program implementation and considered in project design.
- Program design considers and responds to both the individual and the institutional capacity-building required to create sustainable employment for the targeted groups.
- Short-term employment interventions, such as cash for work, which may be used to provide immediate income streams and rebuild assets, must link to a longer term employment strategy that promotes sustainable employment.
- Where applicable, link to existing local and national strategies while applying the do no harm approach (see also *Core Standard 4*).
- Critical cross-cutting themes, including DRR and gender mainstreaming, are integrated into the program design.
- The areas of employment encouraged do not negatively impact the population or environment.

Guidance Notes

 Private-sector partnerships

Partnering with private-sector actors to ensure job sustainability post-intervention is strongly recommended. In addition to ensuring sustainable sources of employment and revenue, private-sector partners can also provide training and support through internships or work-study programs. Private-sector partners that follow international and local labor laws and customs, and are socially responsible, can ensure livelihood stability, while also providing supply to meet market demand.

❷ Short-term interventions

In crisis environments, interventions to promote short-term employment (such as cash for work) are often used as a means of:

- employing vulnerable and/or volatile groups until they can be reabsorbed into the workplace
- injecting cash into the local economy
- restoring/establishing local infrastructure.

These short-term interventions should be leveraged to strengthen participants' potential to fulfil future long-term employment opportunities. Long-term employment can be achieved by identifying, building, and transferring skills from short-term interventions to improve participants' employability in growing industries and/or where there is unmet workforce demand. Additionally, programming should not adversely affect access to other opportunities or divert household resources from productive activities already in place (see also *Sphere Handbook, Food Security – Livelihoods Standard 2: Income and Employment*). Additionally, short-term employment interventions should not weaken or undermine community and/or environmental resources that are important for long-term economic well-being. In the rare instance where interventions are designed to provide employment only in the short term, this limitation needs to be clearly communicated to participants to manage expectations of long-term employment.

 Cautionary tale

> In Myanmar following Cyclone Nargis, aid agencies instituted cash-for-work programs to help villagers rebuild their homes and assets. Wage rates were set higher than market rates to help people recover more quickly. This meant that local farmers could not afford to hire agricultural workers to harvest crops because the project timing and wage rates offered by the humanitarian organizations put them in direct competition with the farmers and made labor too costly.

3. Human capital development

Projects should link with government, educational, and private institutions that already provide vocational training and enhance the quality of services and inclusion for a diverse range of participants. Coordinate with existing state and private training structures to ensure the curriculum is relevant, up to date, and aligned with market needs. Training topics must be continually monitored to fit labor demand as it changes over time and as the context evolves from pre-crisis to post-crisis. Where possible, providing officially accredited trainings and certifications will allow trainees to find and advance in their profession.

Employment programming may involve the hiring of project staff, where skills developed by the beneficiary employees may help them secure future non-NGO employment. As always, staff roles should adhere to decent work standards.

4. Incorporating disaster risk reduction, building back better, and peace building

Employment programs, when appropriate and practical, should integrate other elements into their activities. For example, social cohesion considerations and peace-building indicators will be important in certain contexts. In other areas, reconstruction needs to consider situations (through investigation of appropriate data) that have occurred historically and could potentially occur again in the future. Most areas are sensitive to the impact of natural disasters, and DRR considerations should be included in economic interventions where possible. By implementing strategic policies to build coping capacities, there is increased likelihood of building sustainable employment opportunities and strengthening the resilience of economic systems.

⑤ Context, conflict, and environment-sensitive employment

Employment interventions need to follow international and national laws and standards for decent employment of individuals but must also consider local context including social norms, community tensions, and the environment. As such, programs should focus on the goals of targeted individuals, but also look at the communal impact of employment sectors, including climate-smart programming which avoids highly extractive industries and pollution-inducing trades and encourages employers across sectors to pursue sustainable practices in their industries.

Cautionary tale

Following a disaster that destroys numerous traditional fishing boats, an organization replaces those boats with modern ones. It also provides capacity building to a fisherpersons' cooperative to improve their fishing techniques, but leaves out any discussion of environmental impacts and sustainability. As a result, there is considerable damage to local reefs and overfishing.

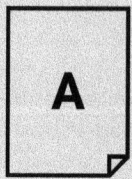

Annex
Market-linked Tools and Frameworks for Assessments

Theme	Tool	Uses	Where to find this tool
Market Analysis Standard	Minimum Standard for Market Analysis (MiSMA) (*Sphere Partner Standard*)	Establishes benchmarks for any market analysis exercise to ensure the quality of humanitarian response and associated contingency plans Best for: market assessments of basic needs and humanitarian relief; crisis focus	Cash Learning Partnership (CaLP): http://cashlearning.org/ > Resources and tools > search Misma
Market analysis	Emergency Market Mapping and Analysis (EMMA)	Useful for understanding market trends for a specific market system in a crisis; can be useful for a wide range of intervention approaches to economic recovery Best for: assessing specific markets for economic recovery, range of intervention options, range of crisis and recovery environments	EMMA Toolkit: http://www.emma-toolkit.org/toolkit
	Pre-Crisis Market Mapping and Analysis (PCMA)	Extra guidance for EMMA created for pre-crisis time periods; can compare information from a PCMA (before crisis) to the results from an EMMA (during crisis) to plan market recovery strategies Best for: assessing specific markets for economic recovery, range of intervention options, cyclical and predictable crises	EMMA Toolkit: http://www.emma-toolkit.org/what-pcma

Theme	Tool	Uses	Where to find this tool
	Rapid Assessment for Markets (RAM)	Resource-light market assessment guidance for crisis contexts; focused on locations rather than specific markets Best for: basic needs and food security (humanitarian relief); 48 hours to 3 weeks post-crisis	International Red Cross and Red Crescent Movement (ICRC): https://www.icrc.org > Resource Center
	Market Analysis Guidance (MAG)	Includes guidance for full program cycle; can cover multiple market systems in target area (less depth than EMMA or MSD) Best for: humanitarian relief and economic recovery; first year following crisis	ICRC: https://www.icrc.org > Resource Center
	Market Information and Food Insecurity Response Analysis (MIFIRA)	For designing market-aware, distribution-based activities for food security Best for: food security (humanitarian relief); crisis focus, can also be applied to recovery	http://barrett.dyson.cornell.edu/MIFIRA/
	Market Assessment Toolkit for Vocational Training Providers and Youth	Resources and activities to assist planning for vocational training and youth programming Best for: economic recovery and employment programs, youth focus	Women's Refugee Commission (WRC): https://www.womensrefugeecommission.org > Resources
	Participatory Market Systems Development (PMSD) approach	How-to on participatory mapping tools and techniques Best for: engaging marginalized actors and participatory planning; recovery contexts	Practical Action: http://www.pmsdroadmap.org/

Theme	Tool	Uses	Where to find this tool
	The Operational Guide for Making Markets Work for the Poor (M4P) Approach	Useful for in-depth analysis to inform intervention strategy; aimed at long-run programming (no specific focus on crisis contexts) Best for: selecting and assessing specific markets, facilitation-based economic recovery; recovery and development contexts	Beam Exchange: https://beamexchange.org/guidance/m4pguide
Political economy analysis	Department for International Development (DfID) Political Economy Analysis: How to Note	Guidance for understanding how power and resources are allocated, can be included with other market analyses Best for: range of interventions; prolonged crisis, recovery, and development	Overseas Development Institute: https://www.odi.org > Publications
	Understanding Policy Change: How to Apply Political Economy Concepts in Practice	Introduces the core mechanisms and logic of political economy and teaches readers to recognize these mechanisms in their daily development-related work Best for: overview of political economy thinking	World Bank: https://www.worldbank.org > Publications
Gender and markets	Gender in Value Chains	Toolkit for integrating a gender lens into a market assessment, such as an EMMA, PCMA, or MSD assessment Best for: ensuring gender is included in market systems assessments	Agri Pro Focus: http://agriprofocus.com/intro > Knowledge Base

Theme	Tool	Uses	Where to find this tool
	Guidelines for Integrating Gender-based Violence Interventions in Humanitarian Action	Assists livelihoods actors and communities affected by crises to coordinate, plan, implement, monitor, and evaluate essential actions for the prevention and mitigation of gender-based violence (GBV) across the livelihoods sector Best for: overall view of incorporating GBV issues across the program cycle	Global Protection Cluster: http://www.globalprotectioncluster.org/ > Tools and Guidance > Essential Protection Guidance and Tools > Gender-Based violence
	CLARA: Cohort Livelihoods and Risk Analysis	Guidance and tools for field practitioners wanting to assess, design, and monitor safe, gender-sensitive livelihoods interventions. Annex 3 of Guidance expands on EMMA tools to cover gender risk analysis. Best for: gathering and analyzing gender data; emergency, can be applied to recovery context	WRC: https://www.womensrefugeecommission.org/ > Resources > Livelihoods
Seed markets	Seed System Security Assessment	Useful for in-depth seed assessments for food security or input-focused programming Best for: seeds; crisis and recovery	Seed System: http://seedsystem.org > Assessment Tools
Household needs assessment	Household Economy Analysis (HEA)	In-depth analysis of household needs; helps to create wealth and livelihood profiles for target groups; thorough household data collection Best for: range of interventions and contexts	Household Economy Approach: http://www.heawebsite.org/about-household-economy-approach
Context monitoring (prices)	MARKit	Price monitoring guidance and templates Best for: ongoing market price monitoring; range of contexts	Catholic Relief Services:http://www.crs.org/ > Research & Publications > Emergency Response and Recovery

Theme	Tool	Uses	Where to find this tool
	WFP: Market Analysis Guidelines	General guidance and tools on market analysis, with greater focus on analyzing price trends over time; for users with some experience in market analysis Best for: ongoing market price monitoring and food market assessments; range of contexts	World Food Programme: http://www.wfp.org/content/market-analysis-guidelines
Monitoring and evaluation	DCED Standard for Results Measurement	Useful for monitoring programs in complex market systems in a standard, systematic way	Donor Committee for Enterprise Development: http://www.enterprise-development.org/measuring-results-the-dced-standard/
Multi-sector assessment	Multi Cluster/Sector Initial Rapid Assessment (MIRA)	Covers the dimensions of status and impact, vulnerabilities and risks, and trends and information gaps in humanitarian crises	Inter-Agency Standing Committee: https://interagencystandingcommittee.org/> Resources
	Post-disaster Needs Assessment / Joint Rapid Needs Assessment	Standardized, comprehensive, post-disaster assessment addressing recovery needs related to infrastructure, shelter, livelihoods, and social and community services	United Nations Development Programme: http://www.undp.org/> Publications > Climate change and disaster risk reduction

Glossary

This glossary provides definitions for the commonly used terminology in the *Minimum Economic Recovery Standards*. These definitions are reflective; the common lexicon is based on widely accepted definitions in work related to economic development, microfinance, enterprise development, livelihoods, market development, agriculture, and food security. Many of these definitions, if not otherwise cited, are adapted from the website of the Microenterprise Development Office at USAID: www.microlinks.org. Definitions cited as *MERS* were developed by contributors to the *Minimum Economic Recovery Standards*.

Access In financial services, access is measured by financial institutions' outreach (in numbers) to micro and small enterprises, with products and services they can use profitably. The definition is similarly applied in enterprise development, in that access is measured by the number of enterprises that can profitably access products and services required for their business, including markets.

Assessment 'Assessment' refers generally to research (both in person and secondary) conducted before and periodically during an economic recovery intervention on market systems, beneficiaries, and surrounding conditions (*MERS*). (See also *Evaluation*.)

Asset protection Most often refers to preventing the sale or consumption of assets by transferring cash or assets (e.g. vouchers, food aid), but may also describe activities to physically protect natural and household assets and ensure access to larger scale or group assets (such as land, water, or group-managed facilities), as well as efforts to ensure that local laws and cultural norms do not endanger people's assets.

Business An occupation, profession, trade, or entity engaged in an economic activity for profit (see also *Enterprise* and *Microenterprise*).

Business linkages Business linkages are mutually beneficial relationships between businesses at the same level of the value chain (horizontal) or at different levels of the chain (vertical) which address constraints at all levels of the chain to support win–win relationships. Business linkages are sometimes also referred to as market linkages. (See also *Horizontal linkages* and *Vertical linkages*.)

Cash-based interventions[1] All programs where cash (or vouchers for goods or services) is directly provided to beneficiaries. In the context of humanitarian assistance, the term refers to the provision of cash or vouchers to individuals, households, or community recipients, not to governments or other state actors. CBI covers all modalities of cash-based assistance, including vouchers, but excludes remittances and microfinance in humanitarian interventions (though microfinance and money transfer institutions may be used for the actual delivery of cash). (See also *Cash transfer*.)

Cash transfer[2] The provision of assistance in the form of money (either physical currency/cash or e-cash) to beneficiaries (individuals, households or communities). Cash transfers as a modality are distinct from vouchers and in-kind assistance.

Child labor[3] Often defined as work that deprives children of their childhood, their potential, and their dignity, and that is harmful to their physical and mental development. It refers to work that is mentally, physically, socially, or morally dangerous and harmful to children, and interferes with their schooling by depriving them of the opportunity to attend school; obliging them to leave school prematurely; or requiring them to attempt to combine school attendance with excessively long and heavy work.

Cooperatives[4] A cooperative is an autonomous association of people united voluntarily to meet their common economic, social, and cultural needs and aspirations through a jointly owned and democratically controlled enterprise. The cooperative model of enterprise can be applied to any business activity. Cooperatives exist in traditional economic sectors, such as agriculture, fisheries, consumer and financial services, housing, and production (workers' cooperatives). They are also found in a wide range of sectors and activities, including car sharing, child care, health and social care, funerals, orchestras and philharmonics, schools, sports, tourism, utilities (e.g. electricity, water, gas), and transport (e.g. taxis, buses). (See also *Group assets* and *Producer groups*.)

Competitiveness The ability of an enterprise or a country to compete successfully, based on price, quality, uniqueness, and/or other socially or environmentally valued standards, with other businesses or countries. Competitiveness is also referred to as sustainable growth in productivity that results in an improved standard of living for average citizens. Achieving and maintaining competitiveness depends on the ability to innovate. Since the competitive advantage of an enterprise depends on the business system and policy environment in which it operates, competitiveness at all levels is interdependent. Success at achieving competitive performance depends not only on enterprises' ability to innovate but also on the performance of both upstream and downstream links in their respective value chains.

Coping strategies Specific efforts that households employ to address disruptions to their sources of income. Common examples of potentially negative coping strategies are reducing daily food intake; consuming cheaper food; reducing household expenditures on necessary items, such as clothing, medical care, and education; and reducing the number of dependents in the household.

Corruption[5] The abuse of entrusted power for private gain, including financial corruption, such as fraud, bribery, and kick-backs. Corruption also encompasses non-financial forms, such as the manipulation or diversion of humanitarian assistance to benefit non-target groups, the allocation of relief resources in exchange for sexual favors, preferential treatment in the assistance of or hiring processes for family members or friends, and the coercion and intimidation of staff or beneficiaries to turn a blind eye to or participate in corruption.

Decent work[6] 'Decent work' sums up the aspirations of people in their working lives. It involves opportunities for work that is productive and delivers a fair income, security in the workplace, and social protection for families; better prospects for personal development and social integration; freedom for people to express their concerns, organize, and participate in the decisions that affect their lives; and equality of opportunity and treatment for all women and men.

Disaster risk reduction[7] The concept and practice of reducing the risk of disaster through systematic efforts to analyze and manage causal factors. It involves reducing exposure to hazards, lessening the vulnerability of people and property, managing land and the environment wisely, and improving preparedness for adverse events.

Economic development As a broad discipline, economic development is defined by different groups based on their target groups and fields of practice. Definitions of the term include: 'Improvements in the efficiency of resource use so the same or greater output of goods and services is produced with smaller throughputs of natural, manufactured, and human capital';[8] and 'the process whereby simple, low-income national economies are transformed into modern industrial economies. Although the term is sometimes used as a synonym for economic growth,[9] generally it is employed to describe a change in a country's economy involving qualitative as well as quantitative improvements.'[10]

Economic growth[11] 'Quantitative change or expansion in a country's economy. Economic growth is conventionally measured as the percentage increase in gross domestic product (GDP) or gross national product (GNP) during one year. An economy can either grow 'extensively' by using more resources (such as physical, human, or natural capital) or 'intensively' by using the same amount of resources more efficiently (productively). Intensive economic growth requires economic development.'

Economic recovery[12] The process of stimulating the growth of an area's local economy through developing markets, strengthening new and existing enterprises, and creating jobs in the private sector and public institutions, including reconstructing needed infrastructure that will allow for trade and commerce to take place in local, national, regional, and international markets. Economic recovery following conflict or disaster should be a transformative process of building back both better and differently, which requires 'a mix of far-reaching economic, institutional, legal and policy reforms' upon which to build self-sustaining development.

Enabling environment An environment of policies, regulations, institutions, and overall economic governance, which allows for economic growth.

Enterprise[13] An enterprise is any entity engaged in an economic activity, irrespective of its legal form. Enterprise development programs focus on self-employed people, family businesses, partnerships, or group businesses (e.g. associations, cooperatives, informal groups) that are regularly engaged in an economic activity. (See also *Microenterprise* for definitions of micro, small, and medium enterprises based on revenue and employee size.)

Enterprise development Enterprise development entails supporting economic activities by individuals and businesses, ranging from self-employment to large commercial operations, whether formal or informal. This can involve directly supporting businesses but also refers to interventions that help an entire market system or value chain function more effectively and in a manner that helps the target beneficiaries to raise their incomes.

Evaluation 'Evaluation' generally refers to post-intervention determinations of a program's performance and effects, for example, outcomes or impact (*MERS*). (See also *Assessment*.)

Exit strategy The plan or strategy for withdrawing relates to withdrawing from subsidizing an intervention, leaving behind sustainable improvements in the private sector.

Facilitator This can be an institution or project that gives indirect support for private-sector development. Rather than providing services directly, a facilitator orchestrates interventions that build local capacity for providing commercial services and/or solutions (to recurrent constraints), preferably through existing providers in the private sector. Services and/or solutions can include access to markets, product development/design, technology access, training, consulting services, links to financial services, improved inputs, and/or advocacy services.

Fair wages[14] These are wage levels and wage-fixing mechanisms that provide a living wage floor for workers, while complying with national wage regulations; ensure proper wage adjustments; and lead to balanced wage developments within a company (with regard to wage disparity, skills, individual and collective performance, and adequate internal communication and collective bargaining on wage issues).

Financial costs Also called 'cost of funds', these are the costs of the funds raised by a microfinance institution to cover its lending. Depending on the context, this may include the interest costs paid to depositors and/or to other financial investors, the rate of inflation, or the opportunity cost of funds received as grants or soft loans from donors, governments, or charitable organizations.

Financial inclusion[15] The provision of a full suite of good-quality financial services to all who are financially capable, by a range of providers.

Financial services In the context of enterprise development, these services include credit, savings, remittances, insurance, leasing, and credit cards (see *Microfinance*). These services are generally targeted to low-income people but may also cover larger enterprises to create employment opportunities for low-income people.

Financial sustainability The degree to which an organization collects sufficient revenues from its services to cover the full costs of its activities, including operating costs, cost of funds (see also *Financial costs*), and expected losses.

Formal sector / formal economy[16] The formal sector or formal economy refers to regulated economic units (e.g. businesses) and workers that are regulated and protected. Put another way, the formal sector comprises economic activities and enterprises that are regulated and/or taxed by the government. (See also *Informal sector / informal economy*.)

Full financial sustainability A situation in which the revenues an organization generates from its clients cover the full (opportunity) costs of its activities, thus allowing it to continue operating at a stable or growing rate without ongoing support from governments, donor agencies, or charitable organizations. When applied to a financial services institution, full financial sustainability requires that the interest and fees the institution collects for its lending equal or exceed the sum of its operational and financial costs, with the latter evaluated on an opportunity-cost basis.

Group assets Assets owned formally or informally by a group of individuals engaged together in a business. Examples of typical group-managed assets are drip/sprinkler irrigation systems, packaging equipment, warehouses, and generators. Group asset transfers tend to be larger in scale (value and size) than individual asset transfers and more concentrated in one location; additional attention before the transfer must therefore be given to evaluating local market impact and implications.

Horizontal linkages Market and non-market interactions and relationships between businesses or individuals performing the same function in a market system (e.g. among multiple wholesalers). Horizontal linkages tend to be longer term, cooperative arrangements among businesses that involve interdependence, trust, and resource pooling in order to jointly accomplish common goals. Both formal and informal horizontal linkages can help reduce transaction costs, create economies of scale, and contribute to the increased efficiency and competitiveness of a sector. Such linkages also facilitate collective learning and risk sharing while increasing the potential for upgrading and innovation.

Indirect interventions Interventions that engage with traders, wholesalers, officials, or policymakers – any party that is not a targeted ultimate beneficiary of the intervention – and which lead to benefits for the ultimate target population. An example of an indirect intervention is the rehabilitation of key infrastructure links in order to increase trade and create jobs for crisis-affected individuals (*MERS*).

Impact An intended change in a high-level program objective, such as enterprise growth or household income. It should be distinguished from intermediate outputs of projects, such as the number of producers organized or the number of trainings provided.

Impact assessment Involves assessing the impact of a project and proving attribution by comparing actual outcomes with a counterfactual – an estimate of what would have happened if the project had not been implemented. The best way of assessing project impact is through a longitudinal sample survey that uses an experimental or quasi-experimental methodology to compare a sample of project participants with a non-participating but otherwise similar control group. Impact is sometimes measured by canvassing participant and/or expert opinion. While such qualitative inquiries can effectively supplement longitudinal surveys, they are not satisfactory substitutes for the longitudinal approach.

Implementing organization In the context of economic recovery, any government or non-government organization that directly provides assistance to microenterprises, or performs other activities intended to improve the environment for microenterprise performance.

Interfirm cooperation Defined as a strategic agreement between two or several businesses involving exchange and/or sharing or co-development of products, technologies, or services; and covering a variety of arrangements between micro, small, medium, and large enterprises, including licensing and sub-contracting relationships, technology, marketing, and other forms of strategic partnering. The primary motivation for this cooperation is to enhance competitive position or market power, decrease transaction cost, and provide access to organizational knowledge and learning. Interfirm cooperation could be an effective mechanism for capacity building in areas such as technology, product and process quality improvements, marketing, and managerial know-how, particularly for micro, small, and medium enterprises. (See also *Business linkages, Cooperatives,* and *Producer groups.*)

Internally displaced persons Internally displaced persons are individuals or groups of people who have been forced or obliged to leave their homes or places of habitual residence, in particular as a result of or in order to avoid the effects of armed conflict, situations of generalized violence, violations of human rights, or natural or human-made disasters, and who have not crossed an internationally recognized state border.

Informal sector / informal economy The informal sector or economy, also called the 'second economy,' refers to work that is not regulated or taxed by the government. It covers a multiplicity of activities and different types of relationship to work and to employment. The informal sector may include the self-employed (in their own activities and family businesses), paid workers in informal enterprises, formal-sector employees with informal second economic activities, unpaid workers in family businesses, casual workers without fixed employers, and sub-contract workers linked to formal or informal enterprises. The vast majority of the world's workers, including the poorest, are in the informal sector.[17] (See also *Formal sector / formal economy.*)

Lean data The process of embedding measurement and data collection in a company's core business activities and using the right tools to rapidly gather information, to improve insights into performance while minimizing time and cost of traditional measurement approaches (*MERS*).

Livelihoods[18] A livelihood comprises the capabilities, assets (including material and social resources), and activities required for a means of living. A livelihood is sustained when it can last through and recover from various stresses and shocks, and preserve or enhance assets and capabilities, while not undermining the natural resources base. UNHCR gives the following definition: 'Livelihoods are activities that allow people to secure the basic necessities of life, such as food, water, shelter and clothing. Engaging in livelihoods activities means acquiring the knowledge, skills, social network, raw materials, and other resources to meet individual or collective needs on a sustainable basis with dignity.'[19]

Market[20] Also called the 'marketplace', the market is any formal or informal structure (not necessarily a physical space) in which buyers and sellers exchange goods, labor, or services for cash or other goods. The word 'market' can simply mean the place in which goods or services are exchanged. Markets are sometimes defined by forces of supply and demand rather than geographical location, e.g. 'imported cereals make up 40 per cent of the market'.

Market analysis[21] The process of assessing and understanding the key features and characteristics of a market system so that predictions can be made about how prices, availability, and access will perform, and decisions made about whether or how to intervene. The term 'market assessment' may also be used to describe this process. (See also *Value chain analysis*.)

Market assessment[22] This is a diagnostic tool that identifies current, recent, and pre-crisis market conditions and trends; supply and demand for goods and services; the characteristics and bottlenecks of supply and value chains; the impacts of crisis on markets; the viability of various income-generating opportunities, occupations and business development; and the extent of access and barriers for crisis-affected populations.

Market-based programming Also called 'market-based intervention', this describes projects that work through or support local markets. It covers all types of engagement with market systems, ranging from actions to deliver a relief intervention to proactively strengthening and catalyzing local market systems or market hubs.

Market chain See also *Sub-sector*, *Value chain*, and *Market system*.

Market development Market development, as defined by The SEEP Network, is a subfield of enterprise-sector development, in which development programs seek to help micro and small enterprises participate in, and benefit more from, the existing and potential markets in which they do business (including input and support markets, as well as final markets). Recognizing that micro and small enterprises do not operate in isolation but rather are part of a larger market, market development involves implementing programs that take market forces and trends into account. This may require that programs work not only at the level of individual small enterprises or households but also with larger enterprises, associations, or government institutions that engage in and influence markets. The ultimate goal of market development programs is to stimulate sustainable economic growth that reduces poverty, primarily by ensuring that small enterprise owners and their employees take part in the growth and reap high rewards.

Market integration A market system is integrated when linkages between local, regional, and national market actors are working well. In an integrated market system, any imbalance of supply and demand in one area is compensated by the relatively easy movement of goods from other nearby and regional markets.

Market linkages See also *Business linkages*, *Value chain analysis*, and *Sub-sector*.

Market rates The prevailing interest rate or fee that is offered in the market reflecting the supply and demand for that service and the cost to deliver that service.

Market system The complex web of people, trading structures, and rules that determines how a particular good or service is produced, accessed, and exchanged. It can be thought of as a network of market actors, supported by various forms of infrastructure and services, interacting within the context of rules and norms that shape their business environment. (See also *Value chain* and *Sub-sector*.)

Medium enterprise See *Microenterprise* and *Small and medium enterprises*.

Microenterprise A very small enterprise owned and operated by poor people, usually in the informal sector, with 10 or fewer workers, including the microentrepreneur and any unpaid family workers. This can include crop production, as long as the activity otherwise meets the definition (USAID). In addition, the World Bank defines a microenterprise as having total assets of up to US$100,000, and total annual sales of up to $100,000 (noting that these numbers, while broadly consistent with those used by most other international financial institutions, depend heavily on choice of this (or any other) definition).[23] The European Commission defines a microenterprise as an enterprise that employs fewer than 10 people with an annual turnover and/or annual balance sheet total that does not exceed €2 million.[24] (See also *Small and medium enterprises*.)

Microenterprise development Any activity undertaken by donors, host-country governments, or non-government organizations to improve the lives of poor people by encouraging the formation and/or improved profitability of micro and small enterprises.

Microentrepreneur Owner and operator of a microenterprise, sometimes an individual who is economically, socially, or educationally disadvantaged, and who usually lacks access to the formal commercial banking system and traditional business development services.

Microfinance The provision of financial services adapted to the needs of low-income people, such as microentrepreneurs, especially the provision of small loans, the acceptance of small savings deposits, and provision of payment services needed by microentrepreneurs and other people who may lack access to mainstream financial services.

Microfinance institution/organization (MFI or MFO) An organization whose activities consist wholly or in significant part of the provision of financial services to microentrepreneurs.

Modality[25] Form of transfer of assets (cash, vouchers, in kind, or combination.)

Operational costs In a financial services context, the portion of a program's costs that covers personnel and other administrative costs, depreciation of fixed assets, and loan losses.

Operational sustainability/self-sufficiency A situation in which an organization generates sufficient revenues from clients to cover all its operational costs.

Opportunity costs The value of a given set of resources in their best alternative use. In the case of economic recovery intervention beneficiaries, this refers to the value of other production they could have done with the assets in question (e.g. other activities they could have done with their time or another crop they could have grown on their land).

Phased approach A programming approach based on the principle that any project may be broken down into a series of steps. Each phase has a clear start point, some well-defined tasks, and a defined end point.

Private sector[26] This comprises entities run by private individuals or groups, usually as a means of enterprise for profit. By contrast, enterprises that are part of the state are part of the public sector, and private, non-profit organizations are regarded as part of the voluntary sector. (Governments, state enterprises, and non-profit organizations are all involved in various market systems, however; for example, as employers, buyers of goods and services, and sometimes as providers of goods and services.)

Producer groups Individuals who are engaged in producing similar products who are organized to achieve economies of scale and production or marketing efficiencies. By (cooperating) organizing into producer groups, micro and small enterprises are often able to: 1) improve their access to and reduce the cost of raw materials through bulk purchasing; 2) increase their efficiency by sharing production skills and resources; 3) enhance the quality and marketability of their products through common production standards and market-driven product specifications; 4) increase access to available financing; 5) obtain critical business services through embedded or fee-for-service mechanisms; and 6) improve their market position by having the quality, quantity, and types of products that multiple buyers demand. (See also *Cooperatives* and *Interfirm cooperation*.)

Productive assets Productive assets are defined as resources that are used to generate income and profit. People can make use of assets in two ways: they can own or directly control them, or they can have access to resources that do not belong to them.

Rapid assessment[27] Rapid assessment is conducted through visits to a number of sites to collect primary (new) data through key informant and group interviews and, sometimes, through questionnaires to a limited number of households. Its purpose is to gain a sufficient understanding of the situation to decide on the type, scale, and timing of response needed, if any. A rapid assessment would normally produce a report within a week when the area is small and/or the population homogeneous, and up to six weeks when the area or population affected is large or heterogeneous.

Remittances[28] Defined as 'income or goods received by individuals or households from other parties who live elsewhere,' remittances typically consist of money that migrants earn abroad and send to their families at home through formal transfers (i.e. banks and transfer operators) or informal transfers (i.e. through social or unofficial channels).

Resilience[29] This is the ability of people, households, communities, countries, and systems to mitigate, adapt to, and recover from shocks and stresses in a manner that reduces chronic vulnerability and facilitates inclusive growth.

Sensitive ecosystems inventories (SEIs)[30] SEIs systematically identify and map at-risk and ecologically fragile terrestrial ecosystems in a given area, with the purpose of encouraging land-use decisions that will ensure the continued integrity of these ecosystems.

Small and medium enterprises (SMEs) The World Bank defines a small enterprise as an enterprise with up to 50 employees, total assets of up to $3 million, and total sales of up to $3 million, while a medium enterprise has up to 300 employees, total assets of up to $15 million, and total annual sales of up to $15 million.[31] The European Commission defines a small enterprise as an enterprise which employs fewer than 50 people and has an annual turnover and/or annual balance sheet total that does not exceed €10 million, while a medium enterprise employs fewer than 250 people, with an annual turnover not exceeding €50 million, and/or with an annual balance sheet total not exceeding €43 million.[32] Also sometimes abbreviated as MSME – micro, small, and medium enterprises (see also *Enterprise* and *Microenterprise*).

Shock Usually sudden, irregular events that significantly affect a household's or enterprise's ability to generate income by regular means. At the level of an economy or market, a shock is an event that disrupts established trading patterns and trends. The effects of a shock will vary among households, enterprises, and markets.

Social entrepreneurship[33] Social entrepreneurs drive social innovation and transformation in various fields, including education, health, environment, and enterprise development. They pursue poverty alleviation goals with entrepreneurial zeal, business methods, and the courage to innovate and overcome traditional practices. A social entrepreneur, similar to a business entrepreneur, builds strong and sustainable organizations, which are set up as either not-for-profits or companies.

Sub-sector[34] A sub-sector can be defined as all the businesses that buy and sell from each other in order to supply a particular set of products or services to final consumers. (See also *Value chain*.)

Subsidy In the context of economic recovery, when all or part of the cost of a good or service is paid by someone other than the user (e.g. an NGO or the government) such that the end user does not pay the full price (*MERS*).

Subsidized credit The provision of loans on the basis of interest rates and fees that fail to cover the full long-run costs of providing those loans.

Sustainability The sustainability of a project impact requires the development of local capacity to address recurring constraints. Recurring value-chain constraints should be addressed with efforts at policy and/or regulatory reform and commercial solutions to supporting (business and financial) services and improved inputs. Moreover, interventions should be temporary and an explicit exit strategy needs to be developed upfront (not at the end of the project) to ensure that an impact is sustainable once project activities end.

Upgrading Refers to a change in mindset, improvements in skills, development of new designs or products based on knowledge of final customers, employment of new technologies, adoption of new functions within a value chain, and other actions that lead to greater competitiveness. Upgrading can involve product development, technology transfer, workforce training, effective backward linkages to suppliers, and the use of information technology to enable businesses to identify and compete in new markets. Organizing micro and small enterprises is often a first step in establishing effective backward linkages to their suppliers.

Value added See also *Upgrading*.

Value chain Describes the full range of activities that is required to bring a product or service from its conception to its end use and beyond, and involves design, production, marketing, distribution, and support to the final consumer. The activities that comprise a value chain can be contained within a single enterprise or divided among different enterprises. Value chain activities can be contained within a single geographical location or spread over wider areas. Global value chains are divided among multiple businesses and spread across wide swaths of geographic space. Evidence shows that global value chains had become much more prevalent and elaborate by the end of the twentieth century. Today, the process of economic development cannot be isolated from these global systems. This means that companies and workers in widely distant locations affect one another more than they have done in the past. Some of these effects are quite straightforward, as when a business from one country establishes a new factory or engineering center in another country. Some are more complex, as when a business in one country contracts with a business in another country to coordinate production in plants owned by yet another company in a third country, and so on (see also *Sub-sector*).

Value chain analysis This focuses on the dynamics of interlinkages within a productive sector, especially the way in which companies and countries are globally integrated. While it includes a description of actors in the value chain and an analysis of constraints along the chain (as do traditional sectoral analyses), a value chain analysis overcomes an important weakness of traditional analysis, which tends to be static and limited to national boundaries. Value chain analysis concentrates on interlinkages and, by doing so, uncovers the dynamic flow of economic, organizational, and coercive activities between producers within different sectors, even on a global scale. (See also *Market analysis*.)

Vertical linkages Links between actors at different levels of a value chain or market system, for example, buyers and sellers. In addition to buying and selling activities, vertical linkages provide for the exchange of knowledge, information, and technical, financial, and business services.

Vulnerable groups Vulnerable populations are defined as those groups of people who are typically excluded, disadvantaged or marginalized based on their economic, environmental, social, or cultural characteristics. While many groups fit this description (e.g. people with disabilities, people living with HIV, and refugees), very poor populations, disadvantaged women, and at-risk youth are the three groups commonly targeted by development programs. It is important to consider that these groups are not homogeneous and not all individuals within these groups are necessarily vulnerable. In particular, there are many women and youth whose social context and individual circumstances do not make them vulnerable.

Notes

1. Cash Learning Partnership (CaLP) (no date) 'Glossary of cash transfer programme terminology', <http://www.cashlearning.org/resources/glossary>.
2. Cash Learning Partnership (CaLP) (no date) 'Glossary of cash transfer programme terminology', <http://www.cashlearning.org/resources/glossary>.
3. International Labour Organization (ILO) (2017) 'Topics', <http://www.ilo.org/global/topics/lang--en/index.htm>.
4. International Co-operative Alliance (ICA), http://ica.coop/en/whats-co-op/co-operative-identity-values-principles
5. Transparency International (2016) <http://www.transparency.org/>.
6. International Labour Organization (ILO) (2017) 'Topics', <http://www.ilo.org/global/topics/lang--en/index.htm>.
7. Sphere Project (2011) 'Glossary', *Humanitarian Charter and Minimum Standards in Humanitarian Response*, <http://www.sphereproject.org/handbook/glossary/?I=D&page=2>.
8. UNESCO, <http://www.unesco.org/education/tlsf/extras/tlsf_glossary.html>
9. Encyclopaedia Britannica, <https://www.britannica.com/topic/economic-growth>
10. Encyclopaedia Britannica, <https://www.britannica.com/topic/economic-development>
11. Soubbotina, Tatyana P. (2004) *Beyond economic growth: an introduction to sustainable development*. WBI learning resources series. Washington DC: World Bank, <http://documents.worldbank.org/curated/en/454041468780615049/Beyond-economic-growth-an-introduction-to-sustainable-development>.
12. Women's Refugee Commission (WRC) (2009) *Building Livelihoods: A Field Manual for Practitioners in Humanitarian Settings*, New York: WRC.

13. European Commission (2003) 'Commission Recommendation of 6 May 2003 concerning the definition of micro, small and medium-sized enterprises (notified under document number C(2003) 1422)', *EUR-Lex,* <http://eur-lex.europa.eu/LexUriServ/LexUriServ.do?uri=CELEX:32003H0361:EN:NOT>.
14. Fair Wage Network (no date) 'Definition of fair wages', <http://www.fair-wage.com/en/fair-wage-approach-menu/definition-of-fair-wages.html>.
15. Cash Learning Partnership (CaLP) (no date) 'Glossary of cash transfer programme terminology', <http://www.cashlearning.org/resources/glossary>.
16. Chen, M.A. (2007) 'Rethinking the informal economy: linkages with the formal economy and the formal regulatory environment', DESA Working Paper, no. 46, ST/ESA/2007/DWP/46, New York: UN Department of Economic and Social Affairs, <http://un.org/esa/desa/papers/2007/wp46_2007.pdf>.
17. Women in Informal Employment, <http://www.wiego.org>, and the ILO (2017) 'Informal economy', <http://ilo.org/global/topics/employment-promotion/informal-economy/lang--en/index.htm>..
18. UNDP, Guidance Note on Recovery, Livelihood, http://www.unisdr.org/files/16771_16771guidancenoteonrecoveryliveliho.pdf
19. United Nations High Commissioner for Refugees (UNHCR) (2014) *Global Strategy for Livelihoods: A UHNCR Strategy 2014–2018*, <http://www.unhcr.org/530f107b6.pdf>.
20. International Rescue Committee (IRC) (2016) *Revised Pre-crisis Market Analysis (PCMA)*, <http://www.emma-toolkit.org/sites/default/files/bundle/PMCA_FINAL_WEB.pdf>.
21. Adapted from Albu, M. (2010) *Emergency Market Mapping and Analysis Toolkit*, Rugby and Oxford: Practical Action Publishing and Oxfam GB.
22. UNDP, Livelihoods and Economic Recovery in Crisis Situations, http://www.undp.org/content/dam/undp/library/crisis%20prevention/20130215_UNDP%20LER_guide.pdf
23. World Bank, <http://documents.worldbank.org/curated/en/819161468766822276/pdf/multi0page.pdf>.
24. European Commission (2003) 'Commission Recommendation of 6 May 2003 concerning the definition of micro, small and medium-sized enterprises (notified under document number C(2003) 1422)', *EUR-Lex,* <http://eur-lex.europa.eu/LexUriServ/LexUriServ.do?uri=CELEX:32003H0361:EN:NOT>.
25. Cash Learning Partnership (CaLP) (no date) 'Glossary of cash transfer programme terminology', <http://www.cashlearning.org/resources/glossary>.
26. Wikipedia (2016b) 'Private sector', <http://en.wikipedia.org/wiki/Private_sector>.

27. Sphere Project (2011) 'Glossary', *Humanitarian Charter and Minimum Standards in Humanitarian Response*, <http://www.sphereproject.org/handbook/glossary/?l=D&page=2>.
28. Women's Refugee Commission (WRC) (2009) *Building Livelihoods: A Field Manual for Practitioners in Humanitarian Settings*, New York: WRC.
29. United States Agency for International Development (USAID) (2015) 'Resilience at USAID', <https://scms.usaid.gov/sites/default/files/documents/1867/06.30.2015%20-%20Resilience%20Fact%20Sheet.pdf>.
30. Ministry of Environment (2016) 'Sensitive ecosystems inventories', British Columbia Government, <http://www.env.gov.bc.ca/sei/>.
31. World Bank, 'Small and Medium Enterprise Development', <http://www2.ifc.org/sme/html/sme_definitions.html>.
32. European Commission (2003) 'Commission Recommendation of 6 May 2003 concerning the definition of micro, small and medium-sized enterprises (notified under document number C(2003) 1422)', *EUR-Lex*, <http://eur-lex.europa.eu/LexUriServ/LexUriServ.do?uri=CELEX:32003H0361:EN:NOT>.
33. Schwab Foundation for Social Entrepreneurship (no date) 'What is a social entrepreneur?', <http://www.schwabfound.org/content/what-social-entrepreneur>.
34. Lusby, F. and Panlibuton, H. (2004) *Promoting Commercially Viable Solutions to Sub-Sector and Enterprise Development Constraints*, Arlington, VA: Action for Enterprise.

Standards Development Task Force

The revision of the *Minimum Economic Recovery Standards* and the production of the third edition would not have been possible without the dedication and hard work of all the individuals who have been involved, in one way or another, in this revision process. We would like to sincerely thank the 150-plus practitioners and leading experts that contributed to this broad consultative process and attended the regional consultation workshops held in Washington DC, Geneva, Dakar, Panama City, New Delhi, Beirut, and London. Acknowledgement should also be given to the expert readers who provided valuable comments on drafts for the *MERS* handbook. Finally, we would like to extend our utmost gratitude to the following people who have played such a pivotal role in this revision process by contributing their time, expertise, and energy to developing the third edition (note that some individuals have since left the organizations listed here; however, they are mentioned to give credit to the organization's commitment):

Technical Working Group Members

Kyhl Amosson, World Vision International
Nicholas Anderson, Save the Children
Ziad Ayoubi, UNHCR
Dina Brick, Catholic Relief Services
Deena Burjojee, Access Alliance LLC
Karri Byrne, Independent Consultant
Oscar Caccavale, World Food Program
Ruth Campbell, ACDI/VOCA
Jerry Cole, Red Rose
Mayada El-Zoghni, CGAP
Alfred Hamadziripi, World Vision International
Nicole Hark, Lutheran World Relief
Shoshana Hecker, Hecker Consulting
Alison Hemberger, Mercy Corps
Christopher Herby, American Red Cross
Christine Knudsen, Sphere Project
Emma Jowett, Independent Consultant
David Leege, Catholic Relief Services

Meredith Maynard, Relief International
Scott Merrill, CARE
Lili Mohiddin, Independent Consultant
Jenny Morgan, The SEEP Network
Jan Morrow, Samaritan's Purse
Stefanie Plant, International Rescue Committee
Zaki Raheem, DAI
Regina Saavedra, UNHCR
Tom Shaw, Catholic Relief Services
Barri Shorey, International Rescue Committee
Eaw Sierzynska, Global Communities
Emily Sloane, International Rescue Committee
Matthew Soursourian, CGAP
Louise Sperling, Catholic Relief Services
Alexa Swift, Mercy Corps
Alexi Taylor-Grosman, Trickle Up

MERS Steering Committee

Ziad Ayoubi, UNHCR
Karri Byrne, Independent Consultant
Alison Hemberger, Mercy Corps
Joseph Mariampillai, Relief International

Scott Merrill, CARE
Tom Shaw, Catholic Relief Services
Julien Schopp, InterAction

Lead Facilitator: Sarah Ward, Independent Consultant

About SEEP

SEEP is a global learning network. We support strategies that create new and better opportunities for vulnerable populations, especially women and the rural poor, to participate in markets and improve the quality of their life.

Founded in 1985, SEEP was a pioneer in the micro-credit movement and helped build the foundation of the financial inclusion efforts of today. In the last three decades, our members have continued to serve as a testing ground for innovative strategies that promote inclusion, develop competitive markets, and enhance the livelihood potential of the world's poor.

SEEP's 118 member organizations are active in more than 170 countries worldwide. They work together and with other stakeholders to mobilize knowledge and foster innovation, creating opportunities for meaningful collaboration and, above all, for scaling impact.

www.ingramcontent.com/pod-product-compliance
Ingram Content Group UK Ltd.
Pitfield, Milton Keynes, MK11 3LW, UK
UKHW021819240426
5401IPUK00004B/44